Ciara Mulcahy

All about me
Displays

JEAN EVANS

AUTHOR JEAN EVANS

EDITOR SUSAN HOWARD

ASSISTANT EDITORS CHRISTINE LEE AND LESLEY SUDLOW

SERIES DESIGNER LYNNE JOESBURY

DESIGNER SARAH ROCK

ILLUSTRATIONS JESSICA STOCKHAM

PHOTOGRAPHS PENNY SHEPHERD

With thanks to the staff and children of Thompson Park Nursery Centre, a Sunderland Social Services Nursery.
With thanks to staff and children of Selby Cottage Childcare Centre, an initiative of Chester le Street District Council, for their help in the creation of some of the displays in this book.

Designed using Adobe Pagemaker

Published by Scholastic Ltd, Villiers House, Clarendon Avenue, Leamington Spa, Warwickshire CV32 5PR

6 7 8 9 0 5 6 7 8 9

British Library Cataloguing-in-Publication Data
A catalogue record for this book is available from the British Library.

ISBN 0439-01638-X

Contents

THEMES ON DISPLAY
for early years

Introduction

Young children learn best through hands-on experience, and this book provides a range of exciting display ideas which are truly interactive, introducing many stimulating and interesting ways of involving the children.

All about me is divided into five chapters, each focusing on one aspect of the theme.

Chapter 1, 'Myself', sets the scene, with displays which encourage children to look more closely at themselves and what their bodies can do. By looking at changes in their development they will become more aware of the passage of time and how their bodies are affected by this. As they explore daily routines they will begin to realize the importance of taking care of their own bodies.

Chapter 2, 'Senses', introduces each of the five senses as the focus of a display. Children are encouraged to observe, taste, feel, smell and listen to a range of stimulating resources and to express their feelings as they do so.

Chapter 3 looks at different types of clothes. Children are asked to consider why certain clothes are worn at various times of the year. They are introduced to clothes worn by babies, special uniforms, clothes from other cultures and different kinds of shoes. Scientific explorations into the properties of materials are made as children test the absorbency of a range of materials, and there are opportunities to create their own fabric designs.

Chapter 4 explores different types of food. There are opportunities for children to examine and taste foods from other cultures, to learn about how food grows, where dairy products originate and where we buy our foods. The chapter concludes with an exciting doll's tea party.

Chapter 5 presents a 'hands-on' look at toys. Children compare their favourite toys and discover how toys have changed over the years. They can try making their own versions of moving toys, kites, puppets and games and a huge shiny robot. The displays in this chapter will provide enjoyment, not only for the children, but also their parents and carers.

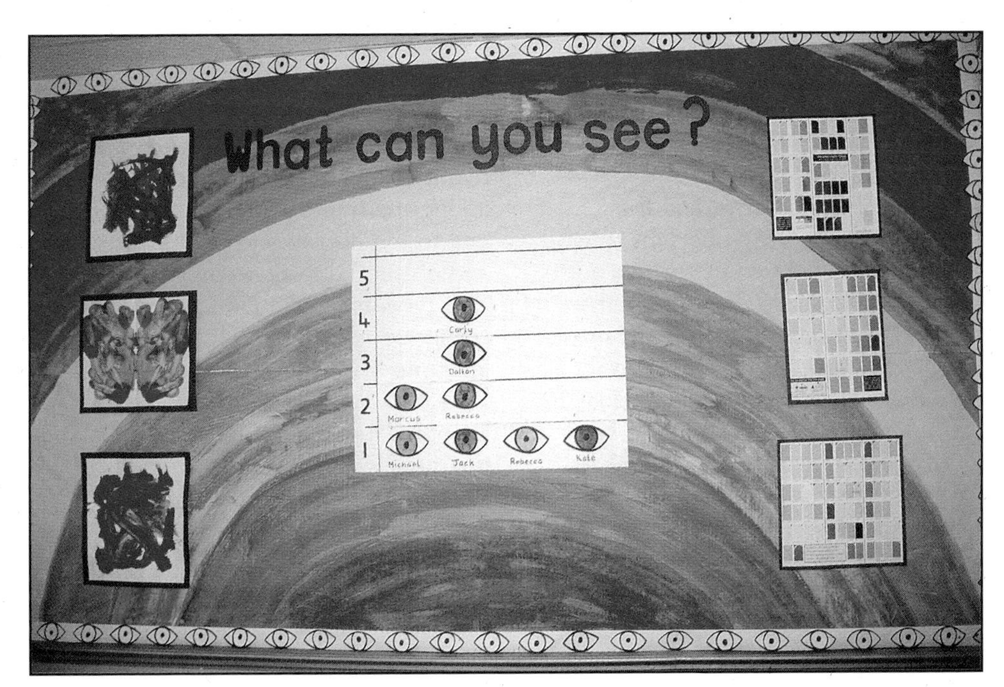

Aims

This book will enable adults working in a variety of pre-school settings to plan, set up and use a range of visually stimulating displays related to the theme 'All about me'. The five chapters can be used individually as the basis for a complete theme or displays from each can be integrated into a larger theme entitled 'All about me'.

Each chapter begins with a stimulus display, followed by five interactive displays and concluding with a table-top display. Each display is closely linked to the six areas of learning and has a specific learning objective related to one of these areas of learning. The aim of the displays in each chapter is to provide a balance across the six areas, with at least one display having a learning objective relating to each learning area.

Planning displays

When planning a theme, and the relevant displays, it is important to work as a whole staff. Begin by making a list of all available spaces for display, their size and position. Photocopy a blank list for repeated use. Refer to your long-term plans to decide on the proposed theme and plan a brainstorming session for suggested displays to relate to each of the six areas of learning.

Pencil the display ideas onto the list of display areas. Make a firm decision about final displays and where they will be situated, ensuring that there is a good balance across the learning areas.

Take each display in turn and make a rough sketch, considering points made in each of the following paragraphs. Make a list of the paragraph headings with a space below and to work through them systematically asking appropriate questions as you do so.

Types of display

There are three types of display in this book.

● Stimulus display

The aim of this display, the first in each chapter, which is inferred in the title, is to 'stimulate' the interest of children and parents in the proposed theme. It could be something that you set up as an introduction to a theme, such as the display 'The five senses' on page 25, which provides a general introduction to the 'Senses' displays in Chapter 2. It might involve children and parents by asking them to bring in contributions to add to the display, such as the 'Toys past and present' display on page 61, which asks parents to bring in toys from their childhood and children to bring in favourite toys from home.

Each stimulus display page includes a list of resources, instructions for making the display, ideas for talking points and suggestions for home links.

● Interactive display

Children learn by experience and exploration and the aim of the interactive displays is to provide them

with the opportunity to do this. They will be able to discover for themselves some of the properties of materials and be able to ask questions about how things work and why things happen. Many of the displays provide sensory experiences and encourage the children to express their feelings and use their imaginations. From these interactive experiences, using resources arranged on tables or floor space, the children will develop their own ideas. The accompanying wall display will stimulate their natural curiosity and extend their knowledge further.

The pages for each interactive display include the same advice on preparation as the stimulus display page and also suggest ways of extending the ideas in the display to promote each of the six areas of learning. These suggestions provide excellent activity ideas when planning a theme.

● Display tables

Table-top displays provide opportunities for further exploration on a particular aspect of a theme, without the accompanying wall display. For example, the 'Match the pairs' table-top display on page 48 provides children with the opportunity to develop the mathematical skill of matching and introduces the concept of pairs. Additional resources enable the table to be converted easily into a 'shoe shop' or to practise 'shoe cleaning' to encourage imaginary play and personal skills.

As well as advice on setting up the table-top display, each page includes further display table ideas based on the same theme, so that learning opportunities can be extended.

Home links and working with parents

Every display includes suggestions for home links. These might involve inviting parents or family members to talk to the children about something related to the focus of the display. For example, the 'What can you see?' display on page 26 suggests inviting an optician or someone who has a visual impairment in to your setting to talk to the children.

Other suggestions include supplying a list of items required so that parents can contribute and inviting them to help with the setting up of the display. Photocopiable sheets are included at the back of the book and these can be sent home with the children so that they can try some activities with their parents.

Encourage parents to be involved with the whole process of creating the displays and to spend time exploring them with their children.

Children

The most important participants in the creation of the displays are the children. Remember to involve them as much as possible at every stage and always let them help to choose colours and

materials. As far as safety allows, let them explore the displays freely and only interact to make suggestions, answer questions and develop their ideas further by extending resources.

Setting up the displays

Preparation of boards

It is essential when taking down old displays that the board is thoroughly cleared of pins, scraps of paper and staples so that there is a smooth surface to begin the next display. All too often, displays are mounted on boards which have half-pulled out staples and left-over drawing pins in them, and this can cause injury to fingers.

Background papers

Plain frieze paper is usually the best form of backing paper, but wallpaper can be used to create a particular effect, for example when creating a display about rooms in a house, or when a certain texture would be appropriate, such as a 'blown vinyl' ploughed field. Wrapping paper can create a very effective background. Children's wrapping paper was successfully used for backgrounds in the 'Party foods' display on page 58 and in the 'Baby clothes' display on page 46.

Colours

Consider the overall effect of colours used in a display at the planning stage. If the display is full of large collage work and labels, such as the 'My body' display on page 14, it is usually best to choose a plain colour for the background so that the displayed work stands out.

When the background forms part of a scene, such as sky and grass, blue and green will be the main colours, but these can be enhanced by sponge-painting different shades of the same colour on top of the paper.

Certain colours are associated with different temperatures. Choose appropriate colours, such as red and orange for 'hot' pictures and blue and white for 'cold' pictures.

Mounting

It is important for children to see that their work is valued, and this can be achieved by the careful mounting of their pictures for display. Ensure that the work is neatly trimmed to remove rough edges and mount on paper to leave a border showing. Mounts can be in direct contrast with background paper, or black so that they stand out. The children's self-portraits in the 'Myself' display on page 18 look particularly effective mounted on paper which matches the title lettering and contrasts with the background.

For really special pictures try double mounting so that there are two borders, with the one nearest the picture matching the display background.

Finished mounted work can be attached directly to the display with a staple gun or stuck to a small box and glued to the display board to create a three-dimensional effect.

Labels, captions and titles

All labelling should be clear and consistent. Young children are just becoming familiar with print as a form of communication and so it is best to stick to simple lower case lettering with upper case only at the start of sentences and for the initial letters of names.

If possible, use a computer to print out lettering to achieve consistency. Make letters large enough to see at a reasonable distance. Alternatively, use stencils or draw around large wooden or plastic letters.

Black lettering stands out, particularly for the title of the display, but sometimes coloured lettering to match the mounts on the pictures is effective. When choosing colours for lettering and mounts ensure that the chosen colour is a direct contrast to the background

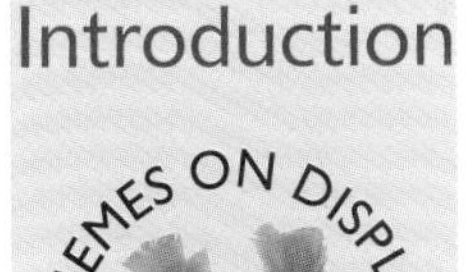

paper so that it is easy to see. Once the lettering is printed it can be cut out and mounted on white or coloured paper, or glued directly to the background.

Borders

An attractive border finishes off a display effectively and draws the eye in to the main components of the display. Try varying the border on each display. Commercially produced borders can easily be obtained from early years catalogues. They are available in a variety of shades and textures and with plain and scalloped edges. In addition, themed borders, such as 'picket fences' or 'icicles', are useful for country and winter scenes.

Home decorating shops often sell odd rolls of border paper cheaply and these can sometimes be related to themes, such as 'toys' and 'farm animals'. It is a good idea to build up a stock of these as they become available.

Odd rolls of wallpaper can be bought cheaply and these can be sliced into border rolls. Textured papers, such as wood chip and blown vinyl, were very effective when creating the 'Feel the animals' display on page 30.

Appropriate wrapping paper can be cut into strips to create unusual borders.

Children enjoy creating their own borders using techniques such as printing. This can be done easily on low displays, but for higher displays the background paper needs to be cut to the shape of the display board beforehand so that the children can print the border before it is attached to the board.

Sponge is ideal to cut shapes from, and potatoes or small plastic dough cutters can also be used. Sponge paint rollers come in various sizes and designs and can be rolled along the paper in wavy lines to create an unusual border.

'Home-made' borders are instantly attractive. The cut-out people surrounding the 'My body' display on page 14 were great fun to make.

Always involve the children in decision making when making borders. The 'flag' border for the 'Food from other cultures' display on page 52 arose from a child's suggestion and was used instead of the original plan for a border of pasta.

Sometimes a display can be framed without the need for a border. In the 'Puppets' display on page 68 fabric is used to create an attractive draped curtain frame around the puppet theatre.

Fixing

The most effective way of fixing the background paper and border to the display board is with a staple gun. However, if this method is used, care must be taken to remove all staples carefully after the display is taken down, preferably using a staple extractor. If you do use large drawing pins to secure paper, make sure that they are pushed well into the board to avoid the danger of children pulling them out.

Once the backing paper is in place, it is best to attach paper and card display features with paste, and thicker items, such as fabric, with PVA glue rather than drawing pins to prevent the ends from curling and to provide a neater finish. Sticky tape loops can also be used for fixing smaller items, such as boxes, to the display.

For those groups without display boards it is useful to use a portable piece of pinboard, but care must to taken to stand this up securely behind a table so that children are not able to pull it over. Heavy items can be hung on display boards by screwing cup hooks securely into the board and using a loop of string.

Three-dimensional displays

Three-dimensional effects are attractive and add interest. They can be created on the display board by gluing some items to small boxes so that they stand out from the rest of the work, or by padding out parts of the display by putting scrunched-up newspaper behind. This technique was used to give the vegetables on the 'Foods that grow underground' display on page 54 an interesting textured 3-D effect.

Table displays can be made more visually stimulating when objects are arranged on different levels. This can be achieved by arranging cardboard boxes, sponge or plastic bricks on the table before draping the fabric over. If the fabric is pinned to the base of the display board it can then be draped across the gap between the wall and table to create a continuous effect.

carrot

Shapes and levels
Create added interest by using different levels and shapes. Have tables of different heights or include floor space. Use circular, square and rectangular tables and put boxes and bricks on them to display items.

Using structural features
Create exciting displays in unusual spaces and make use of structural features, such as window sills, doors, stair rails, pillars and pipes. Twine a long piece of rope with paper leaves attached to a cold water pipe in the corner of a room, or a central pillar, to create a 'beanstalk' for Jack. Use recessed windows to create scenes. Try making an underwater environment by arranging sand and shells on a window sill, hanging fish and green Cellophane weed from above, sticking starfish on the window panes and covering the whole of the front of the recess with blue Cellophane so that the children can look into this stimulating undersea world. Mount pictures on the backs of doors and use stair rails to hang mobiles.

Using awkward areas
Drapes are very effective to cover the more unsightly aspects of some rooms, such as pipes, flaking wallpaper and damp patches. Position a table in front of the drape so that it can be put over it in folds to become incorporated into the display. Awkward corners can be treated in a similar way.

Semi-permanent and moveable displays
Some groups have no permanent display space and all equipment has to be stored away after use. If this is the case, make use of trolleys to arrange items on and hang a portable cardboard display of children's work along the front. Look around for any mobile vertical surfaces, such as the backs of book display racks and home corner screens.

Displays in different rooms
Consider the importance of displays in all the rooms available to you, not just the main activity room used by the children. Create simple displays about personal hygiene in the wash rooms. Cover these with sticky backed plastic to protect them.

Position displays about food, healthy eating, table manners and safe handling of food in the area where children have meals or snacks. Do not have displays in kitchens, other than laminated posters, for health and safety reasons. Remember that displays in staff and parents' rooms are equally important and ensure that they are instantly attractive and of high quality.

Outdoor displays
Think about using fences and walls to create murals related to outdoor topics, such as wildlife or weather changes. Involve local art colleges in the production of these.

Photographs

Remember the importance of photographs as a means of explaining to parents the learning objectives that you propose to cover through planned activities and displays, to help children to recall a sequence of events and as a means of evaluating work and making modifications for future plans. Take photographs of the preparation of each display, including pictures of the children painting and making collage pictures and the staff preparing the background. If the work involved an additional feature, such as the 'sound walk' in the 'What can you hear?' display on page 28, then take photographs of this. The collection of photographs can then be used as an additional display or to make a book. Add appropriate explanations about each stage which will be of interest to either staff, parents or children, according to the purpose of the book.

Artefacts

For every theme it is useful to begin to collect together a box of artefacts for future displays. Add to these by making requests to parents. Hang up a list of suggested items required or send home a letter. Unusual artefacts can sometimes be borrowed. Some towns have museums and multicultural centres which have loan schemes.

Sources of free information and resources

Make appropriate use of local recycling centres and businesses for free or cheap materials. Paper and cardboard offcuts can often be obtained from printers.

Books

Make appropriate use of libraries to extend your book supply. Often librarians will suggest suitable titles for a given theme. Involve the children in making their own books as often as possible and hang these next to the display using plastic cup hooks and string. Use a favourite story as a display stimulus.

Lighting

Consider how the light falling on a display can alter the appearance. Make use of natural light by creating displays in windows using coloured Cellophane and tissue paper to let the light pass through. Underwater scenes and stained glass windows are particularly effective.

Safety

Always make sure that children handle tools and materials safely. Use staple guns away from children and do not let young children handle staples and drawing pins. Ensure that children only work on the display at floor level and do not have to stand on chairs to reach it. Seek advice from the fire department about materials to use for displays in entrance halls, stairs and corridors.

Myself

In this chapter you will find a wealth of stimulating display ideas to help children learn all about themselves, from thinking about how they have changed since they were babies to considering their own feelings and expressions.

Our group

Learning objective: to begin to develop positive relationships with staff and children within the group.

What you need

Bright backing paper; contrasting coloured card; camera; black felt-tipped pens; paper in various flesh colours; scissors; stapler; glue sticks; table; cardboard box; fabric; photograph albums of group activities; books about nursery, such as *Starting School* by Janet and Allan Ahlberg (Puffin) and *Going to Playgroup* by Catherine and Laurence Anholt (Orchard Picture Books); adult and child-sized chair.

What to do

Send home letters asking for clear photographs of the children, or take your own. Take pictures of all of the children taking part in activities, Take photographs of the staff.

Let the children help with the photography and with arranging the activity photographs in the albums and helping to write captions. Cover the albums with paper to match the wall display. Mount the 'face' photographs on card and write name labels.

Cover the display board with bright paper. To make a border, cut circles of different shades of 'skin coloured' paper, approximately 10cms in diameter, and let the children draw happy faces on them. Create the caption 'Welcome to our group'. Attach staff photos to the centre of the display and surround with the children's photos. Add the name labels. Place the table in front of the display and position the box on top. Drape with the fabric and arrange the albums and books on top. Position the chairs nearby.

Talk about

- Look at the names of the children. Do any have the same names? Are they always spelled the same?
- Count the number of boys, girls and staff.
- Discuss the activities in the albums. Which ones are the children's favourites?

Home links

Send home a letter encouraging parents to contribute photographs of their children and inviting them to enjoy the finished display with their children.

My body

Learning objective: to extend vocabulary and to begin to recognize familiar words.

What you need

Light coloured backing paper; card; large sheets of white paper; paint; brushes; glue; scissors; felt-tipped pens; children's clothes; collage materials; drawing pins; wool; table; fabric or paper to cover table; small shallow box; jigsaws; books such as *Find Out About My Body* by Anita Ganeri (BBC); two dolls.

What to do

Ask the children to think about the different parts of their bodies. What do they use their feet for? What do their knees help them to do? Help them to name each part of their body. Tell the children that you are going to make a display which shows the different parts of the body.

Back the display board in light paper and prepare the border by folding and cutting out rows of paper people. Let the children draw faces on them, then attach the rows around the edge of the display.

Choose a boy and a girl and ask them to lie down on a large piece of paper. Draw around them and cut the outlines out. Let the children paint the limbs and faces of the cut-outs and glue on facial features and hair using collage materials, such as buttons, wool and fabric. Glue one outline to each side of the display board leaving enough space at either side to attach labels. Ask the children to choose clothes to dress each cut-out and staple these securely to the display. Make card labels for the body parts and discuss with the children where to stick these on the display. Join the label to the relevant body part with wool, fixing securely at each end with a drawing pin.

Create the caption for the centre of the display in large letters and stick this on.

Make a matching game by drawing an outline of a child on a small piece of card. Discuss with the children where to write the names of the body parts around the outline and draw lines to join the names to the correct body

parts. Make a set of matching word cards and put these in a small shallow box. The children can cover the words on the picture with the correct labels.

Place a table in front of the display board and cover it with fabric or paper. Sit a doll without clothes at each side and arrange games, jigsaws and books in between. Put the matching game in the centre.

Talk about

- Ask the children to name the body parts on the display and find them on their own bodies and the dolls.
- Play the games and encourage the children to use appropriate words to describe the pictures.

Home links

- Send a sheet of rhymes and songs about body parts home such as 'Head and shoulders, knees and toes' and ask parents to sing them with their children.
- Make a simple version of the body parts game for children to play with their parents.

Using the display

Personal, social and emotional development

- Talk to the children about caring for their bodies. Emphasize the importance of keeping clean, having lots of exercise and enjoying healthy food.

Language and literacy

- Learn the vocabulary for the different parts of the body. Enjoy action rhymes using different parts of the body.
- Extend the children's knowledge through books, showing them how to handle them appropriately.
- Encourage children to recognize 'body' words through matching activities and games.

Mathematics

- Count the different parts of the body. How many legs, arms, fingers and toes do we have?
- Introduce mathematical vocabulary as you talk about who is tallest or shortest, who has long hair or short hair and so on.
- Make patterns as you create the border of folded people.
- Sort, match and compare using the games and jigsaws.

Knowledge and understanding of the world

- Explore the features of human beings and introduce appropriate vocabulary for body parts. Discuss any similarities and differences. Compare adults with children, and discuss the differences between babies, children and old people.
- Use the computer to create some of the lettering.

Physical development

- Explore body movements as the children enjoy action rhymes and develop control and co-ordination.
- Encourage manipulative skills using a range of tools and materials as the children collage the figures.

Creative development

- Provide opportunities for children to explore colour, texture, shape, space and form as they work in two dimensions to create the border and matching game, and in three dimensions to create the collage figures.

THEMES ON DISPLAY for early years

Who am I?

Learning objective: to become aware of past and present and the passage of time.

What you need

Low display board; baby gift wrap; dark coloured paper and card; ribbon; baby photographs and recent photographs of children and staff; potato; paint; glue; sticky tape; two photocopies of each photograph (enlarged to A4 size if possible); three boxes; hole punch; sugar paper; string; felt-tipped pens; baby items such as rattles, bottles and toys; books about how babies and children grow such as *Here Come the Babies* by Catherine and Laurence Anholt (Walker Books) and *The Baby's Catalogue* by Janet and Allan Ahlberg (Puffin); stick-on plastic hooks.

What to do

Before beginning the display, send home letters asking for a baby photograph and a recent photograph of each child. Emphasize that simple shots of the child's face will be best. Alternatively, let the children draw pictures of what they think they looked like as babies and take your own photographs of each child.

Invite the group to look at all of the photographs or drawings. Can the children identify themselves? How have they changed since they were babies? What things can they do now that they couldn't do when they were babies?

Make 'baby' books using large sheets of sugar paper. Punch holes down the left-hand side and 'sew' together with string, then attach hanging loops. Inside, include the children's drawings, photographs, pictures of babies and baby items cut from catalogues and wrapping paper.

Cover the display board with the baby gift wrap. Make a printed border by cutting a question mark shape in half a potato and a round dot shape in the other half. Remember to cut a reverse image so that the finished print is the correct way round.

Let the children print question marks around the edge of the display.

Glue the recent photographs onto card, leaving a border so that the children can write their names at the bottom. Write the names of younger children for them. Cut pieces of card at least 1cm larger and attach the baby photographs. Make a handle to lift these cards by attaching ribbon to the bottom of each card with sticky tape. Arrange the recent photographs of staff and children along the display board and cover each one with the appropriate baby picture, attached at the top edge with sticky tape so that the children can lift the flap to reveal the photograph underneath. Use dark-coloured paper to create the title 'Who am I?'.

Cover three boxes with gift wrap. Mount the photocopied photographs onto A4 card to create matching games. Put the baby photographs in one box, recent photographs of children in another and recent staff photographs in the third. Add a label indicating 'babies', 'children' and 'adults'. Cover the table with gift wrap and arrange the boxes of photographs, books and baby items on the table. Attach the plastic hooks to the display board and hang the home-made books from them, or place them at the front of the table.

Talk about

- Look at the photographs and discuss any differences in appearance, such as length and colour of hair, height, wearing spectacles and so on.

Home links

- Send home a letter to parents explaining the aim of the display and asking for appropriate photographs.
- Encourage parents to share the display with their children by playing with the photograph cards and by trying to guess who is under the flaps on the board.

Using the display

Personal, social and emotional development

- Observe and compare what a baby can do with the skills the nursery children have.
- Discuss how babies and children learn to walk, talk and feed themselves.
- Why do babies cry? Invite suggestions from the children.

Language and literacy

- Ask each child to choose an item from the table display, name it and talk about it to the others.
- Encourage name recognition using the display.

Mathematics

- Sort, match, count and compare the photocopied cards.
- Use appropriate mathematical language to describe differences in children and the position of photographs.

Knowledge and understanding of the world

- Talk about past and present events in the children's lives.

Physical development

- Pretend to be babies, children and old people. How would they move about?
- Compare toys designed to develop babies' and children's manipulative skills.
- Select appropriate tools and materials to cut and stick photographs, pictures and drawings to make simple home-made books.

Creative development

- Draw and paint pictures of babies and children to add to the display.

Myself

Learning objective: to explore colour, shape and form in two and three dimensions and to express imaginative ideas and feelings.

What you need

Brightly coloured backing paper; contrasting paper in a darker colour; scalloped border paper; white paper; powder paint in various colours; yoghurt pots; shallow plastic trays; paint brushes; paste; paste brushes; safe mirrors including magnifying mirrors; books about feelings such as *Why Can't I Be Happy All the Time? – Questions Children Ask About Feelings* (Dorling Kindersley) and *Where the Wild Things are* by Maurice Sendak (Picture Lions); games; photocards and jigsaws showing different emotions; puppets or dolls with interchangeable happy and sad faces; table cloth.

What to do

Back the display board with brightly coloured paper. Use a commercially produced wavy border in a contrasting colour or create your own by cutting strips of frieze paper in a wavy pattern. Invite the children to look at their faces in the mirrors and talk about their facial features. Can they name common features? Emphasize the differences in the children's skin, eye and hair colour, then talk about the different colours they will need to use to paint their portraits. Provide a range of different coloured powder paint in cut-down yoghurt pots, water, paintbrushes and plastic trays to mix the paint in. Let older children mix their own colours. Show them how to dip the brush into water and then into the chosen colour, spread a blob of paint on the tray, wash the brush and add another colour until they have mixed their chosen shade. Younger, less experienced children will need adult help.

Give the children sheets of white paper and encourage them to fill the paper with large self portraits. When the paintings are dry, help the children to stick their portraits to paper which matches the border. Cut around the

paper to leave a border around the portraits. Let the children cover the backs of the mounted portraits in glue and pass them to you to stick to the display board. Help the children to make name labels, encouraging them to write their own names if possible. Mount the labels on matching paper and stick them beneath the appropriate portraits. Create the title 'Myself' in coloured paper to match the borders. Cover the table in the cloth and arrange the mirrors, books, games and dolls on the top.

Talk about

- Talk about the common features the children share and look for differences in hair, skin and eye colour.
- How do our facial expressions change with our moods? What makes the children feel happy, sad or angry?

Home links

- Send home instructions for making a paper plate puppet with a happy face on one side and a sad face on the other. Attach the plate to a stick so that the children can turn the face around.
- Invite parents to come and look at the portrait display and play the games with their children.

Using the display

Personal, social and emotional development

- Encourage the children to talk about what makes them smile. Look for different facial expressions in books and games and on the cards. Can the children suggest why the people might be smiling?
- Introduce an awareness of cultural differences.

Language and literacy

- Enjoy the books together and talk about how to handle them correctly.
- Develop name and letter recognition, and encourage the children to write for a purpose by making their own name labels.

Mathematics

- Use appropriate mathematical language as you describe the quantity and position of the portraits. Ask appropriate questions such as 'How many boys are there?'; 'Are there more girls than boys?'; 'Whose paintings are on the top row?'.

Knowledge and understanding of the world

- Ask the children about their families as they look at themselves in the mirrors. Do they look like other family members? Remain sensitive to individual children's circumstances during this activity.
- Ask the children to draw small pictures of things which make them happy and sad. Cut around the pictures and make a chart of 'happy' and 'sad' things.

Physical development

- Extend the idea of how different facial expressions portray moods by asking the children to respond to music with their bodies. Which pieces make them feel like dancing? Do any make them feel sad?
- Invent an angry dance and accompany it with instruments such as cymbals and drums. Create a happy dance accompanied by bells and tambourines.

Creative development

- Explore three-dimensional self-portraits using clay or Plasticine.
- Cover a balloon with strips of paper and glue and allow to dry. Paint the 'faces' with different expressions and add woollen 'hair' keeping the knot in the balloon at the top. Hang the faces up by attaching string to the knot.

Hands and feet

Learning objective: to develop skills using hands and feet.

What you need

Light and dark frieze paper in two contrasting colours such as pink and blue; thick paint in lighter shades; shallow plastic trays; aprons; scissors; glue and spreaders; magazines and catalogues; sticky tape; card; fabric; two shallow boxes; pairs of socks and gloves; selection of unusual gloves such as rubber gloves, gardening gloves and mittens; miniature rotary dryer; pegs.

What to do

Invite the children to think about the different things that they can do with their hands and feet. Make a list of the children's suggestions, and talk about each one.

To start the display cover one half of your display board with dark pink paper and one half with dark blue. Spread a large piece of dark blue paper on the floor. Ask four children to put aprons on. Tip a small quantity of the light blue thick paint into trays and ask the children to make handprints on the paper. Repeat the process to make footprints on dark pink frieze paper using light pink paint. When dry, cut out the prints and use them to make borders around the two sections on the display board, alternating the prints up the centre.

Create a title for each half of the display saying 'This little foot can' and 'This little hand can' using your preferred form of lettering. Cut out a giant hand from light blue paper and a giant foot from light pink paper, and attach them to the backing paper.

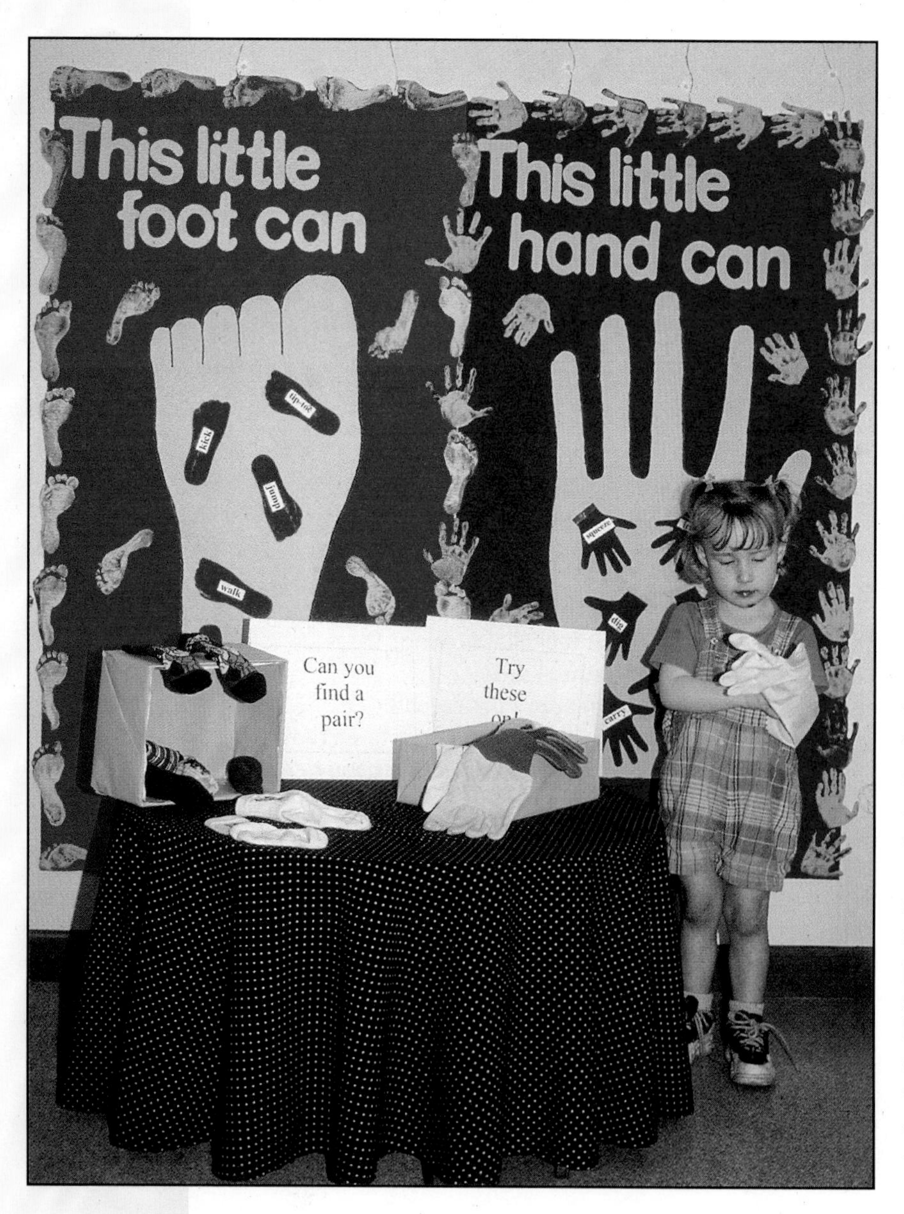

Encourage the children to look through catalogues and magazines for pictures of children doing things with their hands and feet. Ask them to make sure that the pictures are smaller than their own hands or feet. If they are, then they can carefully cut out the pictures. Help them to stick the pictures all over the relevant shape on the display board.

Ask one child to stand barefoot on a piece of dark pink paper. Draw around the child's foot several times and cut out the shapes. Do the same with another child's hand on dark blue paper.

Invite the children to suggest words that describe the actions on the pictures related to feet such as hop, tip-toe or kick. Write these words onto card labels and stick them onto the small feet outlines. Repeat this for the hands.

Use the hands and feet to cover the corresponding pictures on the display; attaching them at the top with sticky tape so that they can be lifted up to reveal the pictures underneath.

Cover the table with the cloth and put a pile of socks and gloves next to two empty boxes so that children can use these for sorting activities. Add some prompts on large pieces of folded card such as 'Can you find a pair?'.

Talk about

- Encourage the children to think about the different parts of our hands and how they do different jobs: we can use our fingers for finger printing and our whole hands to hold a paint brush.

Home links

- Ask parents to bring in contrasting pairs of socks and examples of gloves. Invite them to try the display activities.

Using the display

Personal, social and emotional development

- Arrange hand and foot washing facilities emphasizing the importance of keeping clean.
- Talk about the ways in which hands can be caring; for example by comforting crying babies. Introduce this during role-play with dolls.

Language and literacy

- Talk about the importance of hands for communication. Practise simple sign language and wave coloured flags as signals to 'go' and 'stop'.
- Use hands and feet in action rhymes; such as 'Wind the Bobbin Up' and 'The Grand Old Duke of York'.

Mathematics

- Count fingers in rhymes such as '1, 2, 3, 4, 5' and 'Ten Fat Sausages'.
- Play card games with the children involving matching pairs.

Knowledge and understanding of the world

- Explore unusual hand tools; such as balloon whisks and screwdrivers and talk about how they work. Try to blow up an inflatable paddling pool with a foot pump and a balloon with a balloon pump.

Physical development

- Dramatize movements involving hands and feet; such as climbing ladders. Talk about how some creatures move; such as hopping; jumping and walking sideways, and try to imitate these movements.
- Use small and large balls to develop skills in throwing; catching; kicking and aiming.

Creative development

- Explore sounds made with hands and feet such as clapping; tapping; clicking and stamping.
- Paint finger and toe print pictures.

THEMES ON DISPLAY for early years

Daily routines

Learning objective: to develop an awareness of sequencing and time by exploring everyday routines.

What you need

Bright backing paper; dark paper; white paper; white card; felt-tipped pens; scissors; rubber clock face stamp; glue sticks; sticky tape; A3 sugar paper; drawing materials; catalogues and magazines; four tables at two different heights; fabric; items related to the daily routines of dressing, washing, cleaning teeth, eating, playing and sleeping; books about routines such as *Maisy's Day* by Lucy Cousins (Walker Books) or *Freddie Gets Dressed* by Nicola Smee (Little Orchard Toddler Books).

What to do

Ask parents for photographs of the children involved in daily routines. Place these in a suitable box together with other items for the display such as pictures drawn by the children or cut from magazines and catalogues, and small items such as toothbrushes or plastic knives and forks.

Cover the display board with bright paper. Let the children print clock faces on sheets of white paper, then draw in the two hands with a black felt-tipped pen. Help them to cut the circles out. Use these to make a border for the display. Make the title 'Our routine', using your chosen method of lettering.

Draw six large circles on white card. Mount on sheets of darker paper, leaving a border of about 3cm. Label the circles 'dressing'; 'washing'; 'cleaning teeth'; 'eating'; 'playing' and 'sleeping'. Discuss with the children what to stick on each circle, asking them to choose suitable items from the box and to sort the items into 'routines'. Attach the items to the circles using glue sticks and sticky tape. Bulky items can be pinned on once the circles are in place. Staple the circles to the display. Let the children stick pictures of items associated with daily routines on sheets of A3 paper, adding captions to make posters.

Arrange the tables in two groups at either side of the wall display and

cover them with fabric. Put a set of items for each routine on each surface and two in front of the tables.

● Dressing – include a doll and some clothes. Add a clock to introduce discussion about waking up times.

● Cleaning teeth – include a toothbrush; empty toothpaste tube or box; mug and safety mirror for role-play. Emphasize that this is 'pretend' and children should not put the things in their mouths.

● Washing – arrange a bowl and items such as a sponge; soap; face cloth; hair brush and towel on a table so that children can pretend to get washed.

● Eating – organize a place setting and imitation food on a lower table with a small chair alongside.

● Playing – place a small selection of toys and games in a box on the carpet in front of one of the tables.

● Sleeping – position a doll in a cot in front of the other table. Ensure the doll has night clothes; a teddy and a book.

Talk about

● Try to ensure that staff spend time talking to the children while enjoying the role-play opportunities.

● Talk about the children's own daily routines.

Home links

● Ask parents to contribute items by putting up a suggested list.

● Invite parents to join in with the children's role-play.

Using the display

Personal, social and emotional development

● Encourage personal independence in dressing and hygiene routines during the nursery sessions.

● Introduce cultural differences in routines.

Language and literacy

● Encourage the children to take part in role-play related to daily routines.

● Make a large card circle with Velcro attached at intervals around the edge. Make small cards illustrating daily routines; such as lunch time; story time and outdoor play. Let the children attach these to the larger circle at appropriate times during the day.

Mathematics

● Use appropriate vocabulary related to the passage of time.

● Sort and sequence illustrated cards related to daily routines.

● Complete the photocopiable sheet on page 73 related to daily routines.

Knowledge and understanding of the world

● Develop an awareness of past, present and future by discussing routines and family events.

● Invite someone in to your setting to talk about the daily routines in their work. Make a timeline to record the daily routine of a pet. Ask a parent to talk to the children about the daily routines of a baby.

Physical development

● Mime the movements involved in daily routines such as getting up and eating.

● Collect a selection of brushes used regularly for daily tasks; such as a toothbrush; sweeping brush and hair brush. Explore how these are used.

Creative development

● Use creative techniques such as painting; drawing and model making to communicate ideas about daily routines.

All change

Learning objective: to look for similarities and differences in facial features and to make changes to appearance.

What you need

Two tables; two chairs; fabric; safety mirrors; hairdressing magazines; books about people from different cultures such as *Children Just Like Me* by Barnabas and Anabel Kindersley (Dorling Kindersley) and *Handa's Surprise* by Eileen Browne (Walker Books); masks; spectacle frames; hats; jewellery; hair ornaments; brushes and combs; ribbons; mug tree; shallow display boxes; card; felt-tipped pen.

What to do

Cover the tables with fabric and arrange the mirrors on top. Display the necklaces on the mug tree and hair ornaments in a box on one table, and the masks, spectacle frames and hats on the other. Leave magazines and books in the spaces. Make a sign to display behind the tables saying 'Let's make a change' with a happy and sad face at each end. Let the children explore the items on the table and see how they can make changes to their appearances. Join in with the play and help with fastening of ribbons and jewellery.

Talk about

Discuss similarities in children's facial features and how these can look different by wearing masks; hats; jewellery and different hairstyles. Talk about children's likes and dislikes. Introduce new vocabulary as you decide whether the things on the table can make children look funny, fierce, older, serious, cross, happy or beautiful.

Home links

Ask parents for contributions to the table. Invite them to try changing their own appearances.

Further display table ideas

- Hold a face painting session at a small table organized by a willing parent.
- Make a hat table with a range of hats worn by people in different occupations. Display books about the occupations to extend discussions.
- Have a mask-making table where children can create their own designs on simple pre-cut cardboard masks complete with elastic fastenings. Supply felt-tipped pens; sample masks and pictures or photographs and let the children experiment.
- Have a role-play hairdressing table with appropriate items from the display as well as a bowl, empty shampoo and conditioner bottles, toy hairdryer, tabards and towels.

Senses

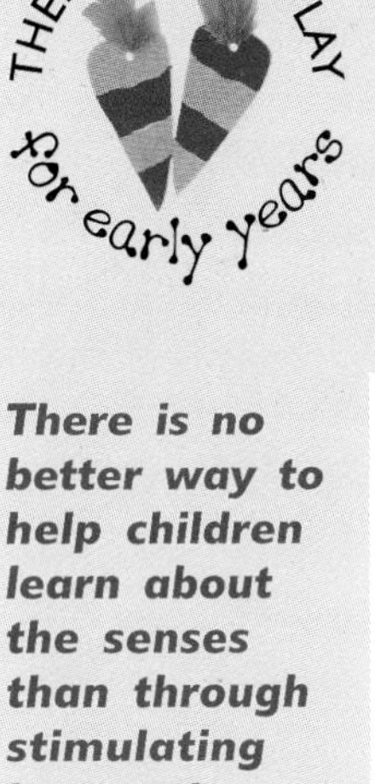

Stimulus display

There is no better way to help children learn about the senses than through stimulating interactive displays, and this chapter includes a range of lively ideas that provide plenty of hands-on opportunities for children to explore and investigate.

The five senses

Learning objective: to become familiar with the names of sensory organs, and to introduce language associated with each sense.

What you need

Backing paper in contrasting colours; black sugar paper; white card; glue; skin coloured paint; fabric scraps; corrugated paper; paintbrushes; pictures and objects related to each sense; table; books such as *The Five Senses* by Sally Hewitt (Franklin Watts).

What to do

Cover the display board in bright paper and add a contrasting border. Cut out five large card circles and glue on items relating to each of the senses.

- Sight: cut two eyes from white card. Add a coloured iris and pupil, and curled black paper eyelashes.
- Hearing and smell: make large papier mâché ear and nose shapes. Paint them using different skin tones.
- Taste: make a mouth shape using fabric scraps and attach a fabric tongue.
- Touch: draw around a child's hands on corrugated paper and cut out.

Write 'sight', 'smell', 'taste', 'touch' and 'hearing' on the appropriate circles. Add the words 'eyes', 'ear', 'nose', 'mouth' and 'hands'. Attach the circles to the board, and arrange items in and around them. Add relevant word labels, such as 'look', 'sniff', 'bitter' or 'shout'.

Place a table in front of the display and arrange the items and books on the table. For health reasons, display just books and pictures of food items.

Talk about

- Ask a child to choose something from the table and to talk to about it.
- Pass around contrasting materials and talk about how they feel. Which is the children's favourite texture?

Home links

- Ask parents with occupations related to senses, such as chefs and opticians, to come and talk to the children.

What can you see?

Learning objective: to record numbers; to understand that people have different needs which should be treated with respect.

What you need
White backing paper; contrasting paper; white card; paint in primary colours and black; sponge; colour charts; plastic trays; fabric; books about sight; samples of Braille; things to look through such as magnifying glasses, camera and binoculars.

What to do
Begin by talking about the sense of sight. Look at each other's eyes. Does everyone have the same colour eyes? Do any of the children wear glasses? Talk about how guide dogs help people with restricted sight.

Discuss the colours that we can see around us, then invite the children to experiment mixing different paint colours together. Put blobs of colour on paper and swirl them around to mix them together. Mount the pictures on bright paper.

Create a graph to show the children's eye colours. Divide a large piece of card into strips and write numbers down the left hand side. Make simple picture cards showing green, grey, blue or brown eyes which fit into the strips on the chart. In turn, invite the children to choose a card which matches the colour of their eyes. Help them to write their name on it, then attach it to the graph.

Spread a large piece of paper on the floor and draw the stripes of a rainbow on it. Let the children paint the stripes. Mount the rainbow on the board. To make a border, cut an eye shape in a piece of sponge and print in black paint on strips of white paper, or simply attach card eye shapes.

Attach the graph to the display, with the children's pictures and the colour charts. Add a title, 'What can you see?'.

Place a table in front of the display and cover it with fabric. Add the things to look through. Include the books and samples of Braille if possible.

Talk about
- Ask a child to choose an item from the table to look through. How does it change the appearance of objects?
- Ask the children to try closing their eyes and describing an object by touch.

Home links
- Invite a parent who is an optician, or someone who has a visual impairment, to come and talk to the children.
- Send home the photocopiable sheet on page 74, inviting families to go for an 'observational walk'.

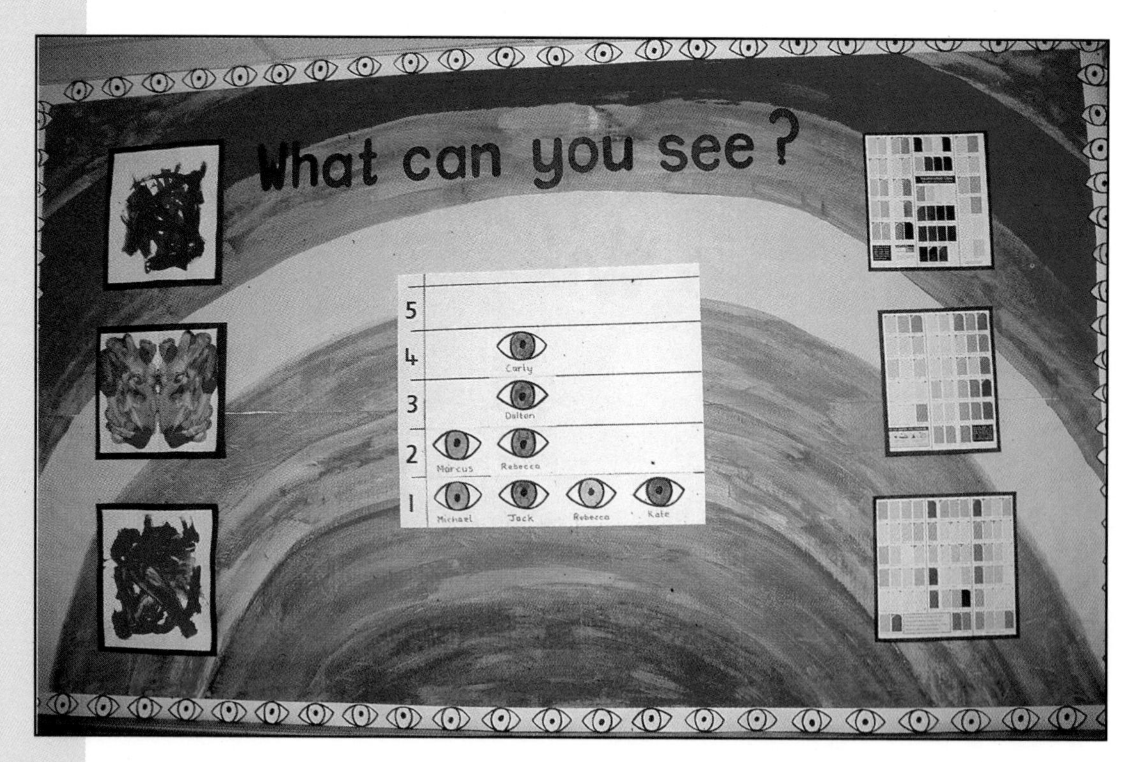

Using the display

Personal, social and emotional development

- Arrange for someone with a guide dog to visit your group and talk about how they manage everyday routines.
- Invite a nurse or optician to talk about how to care for and protect eyes. Remind the children that, even if they are wearing sunglasses, they must never look at the sun.

Language and literacy

- Increase vocabulary by discussing things the children can do with their eyes such as 'wink', 'blink', 'glance', 'flutter', 'close', 'stare', 'narrow', 'open wide'.
- Develop an optician role-play area. Ask a local optician to donate items.
- Increase letter recognition by using an optician's chart during role-play and by playing 'I spy'.

Mathematics

- Look at the chart. What is the most common eye colour? How many children have blue eyes? Are there more children with green eyes or blue eyes?
- Make extra 'eye' cards and play a game, turning them over to find a matching pair.
- How many eyes will two children have? What about three children?

Knowledge and understanding of the world

- Look closely at some examples of living things and discuss whether they have eyes. Where are the eyes on a flat fish? Do plants, worms, caterpillars or ladybirds have eyes? Record your discoveries on a chart.
- Look for similarities and differences in the shape, colour and size of the eyes of different creatures.

Physical development

- Use magnifying glasses to look at tiny objects such as grains of rice. Ask the children to try to pick them up using different tools such as tweezers.
- Try to knock skittles over by rolling balls, emphasizing the need to look carefully at the position of the skittles when rolling the ball.

Creative development

- Make observational drawings of fruits, flowers or vegetables, encouraging the children to look closely at shape, size and colour.
- Experiment with colour mixing using powder paint. Try to match shades or fabric or flower colours.

What can you hear?

Learning objective: to talk about and record observations and to make appropriate use of technology.

What you need

Green backing paper; yellow border paper; collage materials; glue; paint in various colours; maps; bright fabric; musical instruments; boxes with lids filled with items such as conkers, rice and small bells; crêpe paper; items related to hearing such as hearing aids, dictaphone, battery operated cassette recorder; 'sound lotto' game, books such as *Noisy Farm* by Rod Campbell (Puffin) and *Animal Sounds* by Kate Taylor (Walker Books for Sainsbury's).

What to do

Take the children for a short walk around your setting. As you pass different areas, ask the children to listen to the sounds. You may hear traffic, road works, birds or other children. On return, show the children the maps. Tell them that you would like their help to create a map of the route they followed during their walk, which shows all of the sounds they heard. With the children's help and suggestions, draw out a rough plan on a large sheet of paper.

Back the display board with green paper and paint in the path taken or road followed. Add a yellow border. Help the children to make the main features using a range of textured materials and paint techniques, such as trees made from hand prints, bubble wrap ponds, sandpaper shops and wallpaper buildings. Use smaller scraps to create people, animals and birds. Glue the finished features and living things onto the map in the appropriate places. Make captions on bubble shaped paper, such as 'dog barking', 'children shouting', 'ducks quacking' and glue to the display.

Cover the table below with bright fabric. Fill the small boxes with items to create a range of sounds and cover them with crêpe paper. Arrange them on the table with the musical instruments, hearing aids, dictaphone, battery operated cassette recorder and 'sound lotto' game. Stand the books on the table.

Talk about

- Recall the walk and the sounds that the children heard. Try to think of the most

Using the display

Personal, social and emotional development

● Talk about how hearing impaired people communicate using sign language. Teach the children some simple signs.

● Encourage children to respect one another by introducing a 'magic shell' at discussion times. Explain that only the person with the shell can speak, and the others must listen carefully to what is said.

Language and literacy

● Develop children's listening skills using musical instruments. Let them take turns to hide behind a screen and play an instrument while the others try to guess what it is.

● Sing 'Old MacDonald' and compare the sounds that the different animals make.

● Encourage children to associate sounds with patterns in rhymes as you sing favourite songs and nursery rhymes.

Mathematics

● Make comparisons between high and low, loud and quiet, and long and short sounds.

● Play a series of sounds with an instrument and ask children to count how many they hear.

● Make sound patterns by clapping repeat rhythms.

Knowledge and understanding of the world

● Use a cassette recorder to make the same music louder and quieter. Talk about how the volume can be controlled by a knob.

● Show children books and pictures about how ears work. Compare the ears of different creatures.

● Go for a walk and record the sounds that you hear. Play them back indoors and try to identify them.

Physical development

● Use recycled and natural materials to create home-made shakers which make contrasting sounds. Encourage the children to explore and select their own tools and materials.

● Ask the children to perform movements in response to the sound they hear, for example, marching to a drum beat and stopping when the beat stops or running slowly and quickly according to the speed of the beats on a tambourine.

● Invite the children to try to copy rhythms that you play using different musical instruments.

Creative development

● Encourage children to create their own dances in response to music.

● Use a wide range of musical instruments, including home-made, to accompany favourite songs.

appropriate words to describe each sound, for example children screaming, shouting or shrieking; dogs barking, yelping or whining.

● Look at animal pictures and imitate the sounds they make.

● Play the 'sound lotto' game on the table.

Home links

● Send home a sheet with a miniature version of your map. Suggest that families go for 'sound walks' and make their own maps to bring in to add to the display.

● Ask parents and children to make lists of sounds they can hear at home, then bring them in to discuss with the other children.

Feel the animals

Learning objective: to express ideas using a wide range of tools and materials.

What you need

Blue and green frieze paper; different textured wallpapers such as blown vinyl, wood chip and anaglypta; PVA glue; collage materials such as bark, Cellophane, wool and feathers; bubble wrap; string; felt; 'grass effect' material; small boxes without lids; books such as *Look, Touch and Feel With Buster* by Rod Campbell (Campbell Books) and *Touch and Feel* (Dorling Kindersley); sugar paper.

What to do

Invite the children to explore the selection of items with different textures. What does the bark feel like? How about the feathers? Encourage them to describe the textures using words such as 'rough', 'bumpy', 'soft' or 'tickly'. Ask the children to help you make a countryside picture with lots of different textures.

Begin by making 'feely books'. Cut out samples of the different textures and stick them onto sheets of folded sugar paper. Add the children's comments about how the materials feel.

Make this display at the children's height, so that they can touch the different elements. Cover the board with blue and green paper to represent sky and grass. Make a border from samples of textured wallpaper. Talk to the children about what to include in your country scene then involve them in making the different elements. You could include a tree made from bark and shredded Cellophane, a bird made with feathers or a fish made with hologram card. Attach to the display using PVA glue.

Collect samples of each material used and place in separate boxes. Write captions for the display using the children's words associated with the sense of touch, for example, a smooth snail, a shiny snake. Hang up the 'feely books' and stand appropriate books on a small table.

Spread the 'grass effect' material in front of the display and arrange the boxes on the 'grass'. The children can match the samples in the boxes with the textures in the display.

Using the display

Personal, social and emotional development

● Encourage children to talk about their feelings as they handle the materials. Which animals would they like to stroke?

● Talk about the importance of treating living things with care. Use soft toy animals to demonstrate how to pick them up gently and how to groom them.

Language and literacy

● Sit on the 'grass' in front of the display and enjoy an imaginary picnic using items from the home corner.

● Look at the books and enjoy a story together on the 'grass'. Emphasize appropriate handling of the books and encourage children to point to words and letters and turn pages from right to left.

Mathematics

● Sort the samples into rough and smooth, hard and soft, shiny and dull, patterned and plain.

● Discuss the position of the items on the wall display, for example, in the water, next to the sheep or under the tree. Ask the children to point to the different items as you explain where they are.

● Sit on the 'grass' as you sing appropriate 'countryside' number rhymes such as 'Five Little Speckled Frogs' or 'Five Little Ducks'.

Knowledge and understanding of the world

● Talk about visits to the country that the children have made in the past.

● Discuss why animals have fur, feathers, wings, shells, webbed feet and so on.

Physical development

● In a large, clear space dramatize a visit to the countryside. Make large and small body movements and use 'feely' words as you pretend to climb stony hillsides, cross rough wooden bridges, splash in cool, sparkling water and crawl through prickly hedges. End the visit on the 'grass' in front of the display for action songs and rhymes.

● Create smooth patterns with shaving foam on a table and gradually add textures using wood shavings, seeds and pasta. Discuss how the different textures you create feel.

Creative development

● Make three-dimensional models of some of the animals on the display using recycled materials and the sample materials.

● Hide the materials by putting lids on the boxes. Cut holes in the lids and ask the children to put their hands inside to guess the materials by feeling them.

Talk about

● Discuss the different animals in the display. What would the sheep's coat feel like? Would it feel the same as the scales on the fish? Can the children describe how the different textures might feel?

● Introduce opposites such as hard and soft, rough and smooth, dull and shiny into your conversations.

Home links

● Put up a list of things needed for the display and ask parents and carers for contributions.

● Invite parents to help with making the home-made books and to organize the matching activity afterwards.

THEMES ON DISPLAY for early years

Tasty foods

Learning objective: to confidently express personal preferences.

What you need

Purple backing paper; corrugated border paper; paper plates; food magazines; scissors; stapler; glue; food packets; pale blue fabric; paint; plastic food and salt dough models of food; plastic drink bottles; books such as *Sam's Sandwich* and *Sam's Snack* by David Pelham (Jonathan Cape); healthy eating leaflets; small bowls of food to taste such as apple, banana and cucumber slices, raisins and tiny sandwiches (check for allergies to all of the foods used).

What to do

Talk about favourite meals. Do a survey to find the group's favourite food. What is the least favourite? Invite the children to paint and draw pictures of foods that they either like or dislike. Add captions such as 'Ben does not like to eat carrots' or 'Jane loves crunchy apples'.

Cover the display board with the purple paper and add the border. Give each child a paper plate and ask them to look through the food magazines to find pictures of foods that they like. Help them to cut the pictures out, and then glue them onto their paper plate to create a 'meal'. Help the children to write their names on the plates, then staple them around the display. Write the captions 'Food we like to taste' and 'Food we dislike' and attach them to each side of the display, separating the two sections with a line of paper plates. Put the children's pictures and paintings together with sample packets of the foods that they like and dislike on the appropriate sections of the display.

Cover a surface in front of the display with pale blue fabric. Arrange plastic plates of salt dough and imitation food, packets, plastic bottles, healthy eating leaflets and books on the surface.

Add a small tasting table with small bowls of 'tasters' such as slices of apple and banana, raisins, orange slices.

Talk about

- Introduce appropriate vocabulary associated with taste, such as 'sweet', 'sour', 'bitter' and 'salty' and suggest examples of each.
- Compare the taste of raw and cooked foods such as cheese, carrots and apple. Which do the children prefer?

Home links

- Before beginning the display, find out about any food allergies from parents.
- Ask parents and carers to come in to help with food preparation and tasting sessions.

Using the display

Personal, social and emotional development

- Let the children help to prepare samples for the tasting table emphasizing the importance of hygiene routines when handling food and personal safety when handling cutting tools.
- Provide samples of traditional food from a different culture on the tasting table. You could perhaps relate this to a relevant festival, such as Hanukkah or Chinese New Year.

Language and literacy

- Set up two small tables near the display as a café. Encourage the children to use imitation food as they role-play being waiter/waitress or customer.
- Introduce a low table to display recipe books and choose a recipe to follow from one of the books.

Mathematics

- Display some vegetables in a rack near the display. Invite the children to count, sort, match and compare the vegetables according to colour, shape and size. Cut them into smaller chunks and count these. Introduce simple addition and subtraction as you compare the number of pieces.
- Collect empty food containers, such as boxes and plastic bottles, and let the children play with them in the sand tray. Present simple problems involving filling and emptying containers and finding out how many small containers are needed to fill a large one.

Knowledge and understanding of the world

- Talk about the origins of the foods on the display. Did they grow or were they manufactured? Can you discover the country of origin? How? Look at different forms of foods, such as fresh, tinned, dried and frozen potatoes and make comparisons.
- Try using different kitchen utensils such as a fork, spoon, whisk and masher to mash cooked potatoes. Which is most effective?

Physical development

- Make a fruit salad using a variety of tools, such as blunt knives, forks, scissors, plastic graters and juice squeezers.
- Try eating rice using chopsticks, forks and spoons. Which is the most successful method?

Creative development

- Try to identify well-known foods by touching, smelling and tasting. Talk about favourite smells and tastes. Compare the rough, hairy skin of a kiwi fruit with that of a smooth, shiny apple.
- Dramatize the story of 'The Enormous Turnip', then try making turnip soup.

Smells

Learning objective: to respond in a variety of ways to smells.

What you need

Purple backing paper; yellow and skin coloured paper; pictures of things with distinctive smells; strong smelling fruit and vegetables for printing such as lemons and onions; thick paint in various colours; plastic cups; mesh; elastic bands; items to smell – herbs, spices, lemon, vase of flowers, herb plant, pot pourri and lavender bag; books such as *Harry the Dirty Dog* by Gene Zion (Red Fox), *My First Body Book* (Dorling Kindersley), *The Human Body* (*Eyewitness Explorers* series, Dorling Kindersley); plain fabric; cardboard boxes or wood blocks.

What to do

Talk with the children about smells that they like or dislike. Cut the fruit and vegetables in half and let the children use them to print pictures to represent their most and least favourite smells.

Cover the display board with the backing paper. Create nose shapes out of the skin coloured paper and use these as a border. Continue the noses down the centre of the board to make a division. Using the yellow paper, create the captions 'Smells we like' and 'Smells we dislike' and glue these to the two separate sections of the board. Mount the pictures, children's drawings and prints on yellow paper and stick them to the appropriate display section.

Place the table in front of the display board and arrange the cardboard boxes or wood blocks on top to make surfaces of different heights. Cover with the fabric. Write 'Guess the smell' on a sheet of card and stand at the back of the table. Fill the plastic cups with samples of herbs, spices, lemon slices and so on and cover them with mesh, secured underneath with an elastic band. Put the cups on the table.

Arrange the vase of flowers, herb plant, pot pourri and lavender bag on the raised areas of the table and stand the books beside them.

Talk about

● Encourage the children to describe what the smells remind them of as they sniff the cartons. (For safety reasons, do not sniff dry powders before they are covered by mesh.)

● Talk about smells in the countryside and in the town. Are there differences? What does the seaside smell like? Walk outside straight after a rain shower and talk about the smell.

Home links

● Make a 'smelly' present, such as a lavender bag or a pomander, or make some perfume using rose petals.

● Arrange a baking activity. Encourage parents to help with the activity, discussing the different smells with the children.

● Ask parents to take their children on a 'smelling trip', noting the smells in shops, such as a baker's or a pet shop.

Using the display

Personal, social and emotional development

● Discuss the importance of personal hygiene to keep ourselves smelling clean.

● Explore strong smelling foods, such as onions, and then smell their hands. Try using different types of soap to remove the smell. (Check for allergies.)

Language and literacy

● Make home-made books about favourite and unpleasant smells. Encourage the children to draw pictures and copy appropriate words to communicate their ideas.

● Plant herbs in individual pots. Help the children to write their name and the type of herb on a label to stick on the pot.

Mathematics

● Print 'nose' patterns using potato or sponge nose shapes. Use two colours and create alternating patterns.

Knowledge and understanding of the world

● Investigate noses. Do all animals have noses? Are they the same shape? Talk about how dogs find things by sniffing.

● Pour milk into small beakers and smell each one. Add a different powder, such as milk shake mix, drinking chocolate and coffee, to each one, mix, then smell them again. What do they smell of now?

Creative development

● Make pomanders by sticking cloves into oranges and sprinkling them with cinnamon. Tie ribbons around them and hang them up.

● Make perfumed presents following the instructions on the photocopiable sheet on page 75.

● Create nose shapes from clay, dough and Plasticine. Attach the noses to paper plates, then add collage materials to make faces.

Let's make sounds

Learning objective: to show the ability to listen and express imagination through music.

What you need
A selection of musical instruments; home-made shakers; battery operated cassette recorder; song and music tapes; coloured card; sticky-backed plastic in a contrasting colour; scissors; pencil; sticky tape.

What to do
Cut a piece of card the size of the table top and attach to the table with sticky tape. Draw around the instruments on the sticky backed plastic and cut out templates. Stick to the card on the table, making sure that they are well spaced out. Display the instruments on their templates.

Encourage the children to explore the sounds that they can make with the instruments. How can they make the sounds louder and softer? Show them how to operate the cassette player if necessary. Accompany the music or join in the songs.

March around the room playing a band. Invent dances using the instruments. Can the children make up angry, happy and sad dances? Encourage them to return the instruments to their templates after use.

Talk about
- Discuss the different sounds that can be made using one instrument. Does it make a difference if you bang the drum with a wooden beater or a soft beater?
- Talk about favourite songs and music. How does the music make the children feel? Find 'happy', 'sad' and 'angry' examples.

Home links
- Let the children take home a shaker that they have made.
- Send home a suggestion sheet about how to encourage children to listen to music. Include the words of a few favourite songs.

Further display table ideas
- Set up a recording studio on a table. Include a cassette recorder, microphone, headphones and karaoke tapes.
- Make a 'shaker table'. Include identical looking shakers filled with things which will create different sounds. Try to guess the contents of each one.
- Have a table of tuned instruments such as a xylophone, chime bars and bells. Talk about the different sounds and introduce the idea of pitch by talking about higher and lower sounds.
- Create a listening table with stimulating items such as a large shell, tiny bell, coconut shells, sandpaper blocks and a rainmaker. Talk about which items are the children's favourites and why.

Clothes

THEMES ON DISPLAY for early years

Stimulus display

This chapter explores clothes, with exciting display ideas to investigate clothes for different seasons, special uniforms and clothes from other cultures, as well as opportunities to investigate the materials that clothes are made from.

Seasonal clothes

Learning objective: to question why we need to wear different clothes in summer and winter.

What you need

Blue, black and green backing paper; green tissue paper; paint; brushes; glue; collage materials; children's summer and winter clothes and accessories; two tables; blue and yellow fabric; two dolls; winter and summer dolls' clothes; blue and a yellow container; beach towel; summer and winter books such as *Maisy's Mix and Match Mousewear* by Lucy Cousins (Walker Books).

What to do

Attach black backing paper to one side of the display board to look like a winter sky. On the other side attach blue and green to represent grass and sky. Add tissue paper grass to the summer side. Paint a snowy base on the winter side. Make snowflake and sun borders by cutting and painting squares of paper.

Ask two children to lie down on sheets of paper, then draw around their outlines. Let the children paint the figures and add features and hair using collage materials. Glue one outline to either side of the display. Attach real clothes, dressing one for summer and one for winter. Add boots and sandals made from paper or fabric.

Make labels for the clothing and glue them around the outlines, joining to the clothes with wool. Add a suitable title.

Cover one table with blue fabric and one with yellow. Place the blue table on the winter side and the yellow table on the summer side, with the appropriate accessories and books on each. Put the boxes of clothes on the floor in front.

Discuss the display with the children and compare the two different sides.

Talk about

- Ask a child to choose something from one of the tables to talk about. Repeat until all the items have been discussed.
- Discuss the dolls' clothes. Which ones are suitable for hot or cold weather?

Home links

- Ask a parent to bring in a baby with a set of summer and winter clothes so that the children can handle the clothes and compare them.

Clothes from other cultures

Learning objective: to raise awareness of different cultures.

What you need

Bright backing paper; card; white and coloured paper; paint in various colours; felt-tipped pens; glue; scissors; paint and glue brushes; shiny collage materials; wool; magazines; clothing from two cultures; fabric from the chosen cultures; samples of food, utensils, dolls and other artefacts relating to each culture; relevant books such as those from the Wayland series entitled *A Taste of... China* and the Dorling Kindersley series *Children Just Like Me.*

What to do

Choose two different cultures to be the focus of your display and collect clothes and objects related to these. Show the children the items. Can they name them? Compare the clothes with those that the children are wearing. How are they different? Tell the children that you would like them to help you to make a display which shows paper models dressed in clothes from other cultures.

Cover the display board with bright backing paper. Make a border by painting black letters from a script such as Chinese onto white paper. Draw around the outlines of two children on large sheets of white paper. Paint the face, arms and legs of each outline, and make woollen hair, using appropriate colours. Glue one outline to each side of the display, and divide with a strip of paper down the centre.

Ask the children to help you choose appropriate clothes to staple to the outlines. Involve another group in cutting out suitable footwear from fabric or paper. Make jewellery or hair ornaments using shiny collage materials.

Encourage children to draw and paint people in the traditional dress of your chosen cultures and mount these on contrasting paper. Collect photographs and cut out pictures from magazines and mount these. Arrange everything around the display figures.

Add other items such as fans and chopsticks. Choose suitable names for the two children on the display and make a caption for each one.

Using the display

Personal, social and emotional development

● Talk about some of the traditions of the cultures, referring to the books and photographs. Compare these traditions with those of the cultures of the children in your group.

● Develop concentration and turn taking skills by playing 'Kim's game', removing one object at a time and asking the children which one is missing.

Language and literacy

● Introduce new vocabulary associated with the items on display, such as 'sari' and 'wok'. Write the words and draw pictures of the items on pieces of card. Play games matching words to pictures, words to words or pictures to pictures according to your children's ability.

● Encourage the children to use and enjoy the books on display by ensuring that there is somewhere comfortable nearby for them to sit and look at the books by themselves.

Mathematics

● Use mathematical language as you draw around the two children. Who is taller? Where on the board are the outlines positioned? Count the number of items on each table and talk about which table has the most.

● Have a tasting session of foods from one culture. Record the children's favourite food by ticking squares or writing names in a grid. Talk about which is the most and least popular food.

Knowledge and understanding of the world

● Let the children select pieces of scrap fabric and appropriate tools and use their developing skills in cutting and joining to make collage pictures based on the figures on the display.

● Talk about the different textures and appearances of the materials on the display.

Physical development

● Create bracelets and necklaces from strips of paper by colouring them, cutting them out and joining the two ends together.

● Encourage the children to try to copy some of the letter shapes used for the border.

Creative development

● Plan activities based on items in the display, such as making fans, picking up rice with chopsticks and printing patterns on fabric.

Drape two tables with fabric such as sari or kimono material. Arrange your collection of artefacts, food, utensils and books on the relevant table.

Talk about

● Discuss each side of the display in turn. Ask children to choose an item from a table and talk about it. Focus on only one table during each discussion.

● Discuss the different foods on display. Do the children recognize any of them. Have they ever tried them?

Home links

● If possible, ask a parent from another culture to come and talk about their traditions and bring in some items for the children to handle.

● Invite parents to help with the preparation of the display and activities.

Investigating materials

Learning objective: to compare the properties of the materials used to make clothes.

What you need

Blue backing paper; blue and white paint; yellow and white paper; cotton wool; washing line; pegs; two plastic cup hooks; baby or dolls' clothes made of cotton, waterproof material, towelling, wool, fleecy material; samples of some of the materials; pegs; felt-tipped pens; glue; plastic sheeting; baby bath; four buckets; rubber bands; watering can.

What to do

Begin by talking to the children about the different clothes that they wear at different times. Would they wear the same clothes on a sunny day that they would wear on a rainy day? Why not? Talk about what makes some fabrics more suitable than others for different purposes. Tell the children that they are going to investigate the different materials that are used to make clothes.

Use a display board which is low enough for the children to reach. Sponge print blue backing paper with different shades of blue and white paint to represent the sky. Attach it to the display board, then add cotton wool clouds and a yellow sun.

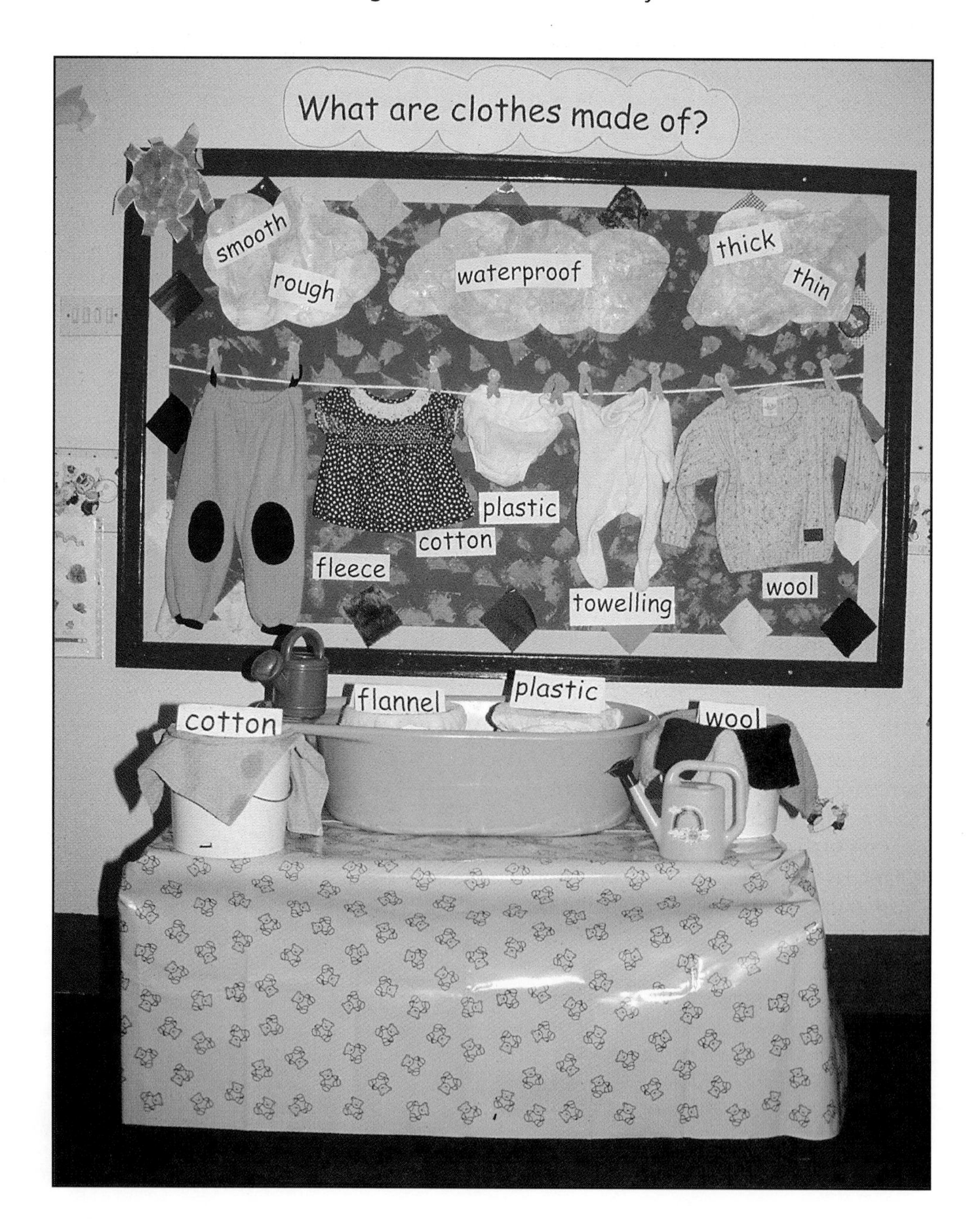

Attach a strip of contrasting paper to make a border and attach squares of different types of fabric such as cotton, towelling, PVC and velour. Attach a plastic cup hook to the frame at each side of the display and fasten a piece of washing line between the hooks. Let the children peg a selection of baby or dolls' clothes to the line ensuring that the range of materials includes towelling, plastic, wool and cotton.

Make labels for the clothes and glue them below the appropriate items on the display. Talk to the children about how the materials feel and write words to describe them, such as 'rough', 'smooth', 'thick' and 'thin'. Stick these words onto the clouds on the display. Make the caption 'What are clothes made of?'

Spread plastic sheeting over a low table and stand a baby bath on it. Cover the top of four buckets with samples of plastic, flannel, cotton and wool, securing them with a large rubber band or string. Stand the buckets in the bath and place a watering can containing water on the table.

Talk about

- Discuss the different kinds of materials the clothes are made of. Touch them and describe how they feel.
- Talk about what happens to the water when it is poured onto the surface of the buckets. Do any of the materials stop the water passing through? Which ones let the water pass through more quickly? Can the children suggest why?

Home links

- Invite parents to contribute to the display and to help with the activity.
- Ask parents to let the children help to sort the clothes at home before they are washed and to talk to them about the materials they are made of.

Using the display

Personal, social and emotional development

- Talk about the importance of keeping clean, washing hands and changing clothes regularly.
- Develop concentration skills and encourage children to use initiative as they choose catalogue pictures of clothing suitable for a rainy day, cut these out and create collage pictures.

Language and literacy

- Introduce the words 'waterproof' and 'absorbs' as the children work on the activity.
- Read the story *The Wind Blew* by Pat Hutchins (Red Fox).

Mathematics

- Count the items on the line. How many pegs are needed for each item? How many altogether?
- Make number cards and attach in order to the clothes on the line.

Knowledge and understanding of the world

- Make a chart to show which materials are waterproof and which soak up water by sticking small pieces of the materials onto squares on a sheet of card.
- Make a rain hat for a doll using suitable material. Pour water over it. Does the doll's head remain dry?

Physical development

- Dramatize the actions of hanging out washing, carrying a heavy basket, and attaching the clothes to the line.

Designs and patterns

Learning objective: to copy and create patterns and designs suitable for fabric.

What you need

Bright backing paper; black sugar paper; white paper; scissors; glue; sponge rollers; powder paints; mix resistant paint (available from early years catalogues); string; cold water dye; white cotton sheeting; marbles; rubber bands; selection of fabrics with different designs including repeating patterns; pattern-making resources such as beads, bricks and kaleidoscopes.

What to do

Gather together to investigate the patterned fabrics. Look at the shapes and colours used to make the patterns. How do the children think some of the patterns were created? Talk about dyeing fabric to produce different colours and printing shapes and patterns.

Invite the children to use different techniques to make their own patterns based on those that they have seen in the fabric.

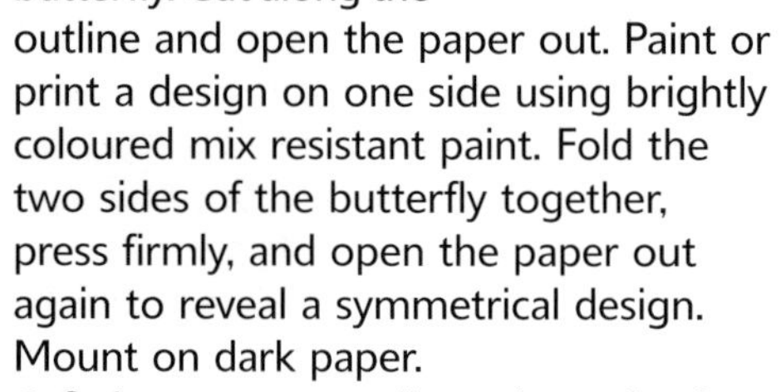

- Symmetrical butterflies – fold a sheet of paper in half and draw the outline of half a butterfly. Cut along the outline and open the paper out. Paint or print a design on one side using brightly coloured mix resistant paint. Fold the two sides of the butterfly together, press firmly, and open the paper out again to reveal a symmetrical design. Mount on dark paper.
- String patterns – dip a piece of string into thick paint and arrange it in a curly pattern on one half of a square piece of paper, leaving an end overlapping the edge of the paper. Fold the other half of the paper over. Press down on the string with one hand and pull it out with the other. Open the paper to reveal the pattern. Mount on square paper.
- Tie dyeing – use large squares of white cotton sheeting. Put a marble on the material and twist the cloth around the marble, securing with a rubber band. Repeat with several marbles. Mix

the dye in a bucket in cold water following the directions given. Drop in the sheets of material with the marbles attached. Leave as instructed to allow the dye to penetrate the material. Take out the marbles and rinse the squares in cold water. Allow to dry. Cut diagonally to make triangles and mount on triangles of bright backing paper.

Prepare the display background by attaching bright backing paper. Let the children use sponge rollers to print strips of paper, then cut into a wavy pattern to make a border.

Mount the tie-dyed triangles on a large square of black sugar paper. Staple this to the centre of the display board. Attach the children's string paintings on either side, with the butterflies on the outside edges. Create labels for each patterning technique.

Place a table in front of the display and cover it with fabric. Arrange resources related to pattern making activities on top, such as beads and cubes with pattern cards to copy, flat wooden shapes, coloured bricks and kaleidoscopes.

Talk about

- Look for the patterns and designs in the pictures the children have made. Which ones do they like best?
- Discuss the different designs in the children's clothes. Are there any repeating patterns?
- Talk about symmetry as you make the butterfly pictures.

Home links

- Ask parents to help their child to investigate clothes with patterns and repeat designs at home.
- Find out whether any parents make their own clothes. Ask them to come in to your setting to show samples of fabric and clothes patterns to the children and to talk about how they use these to create a garment.

Using the display

Personal, social and emotional development

- Look at the designs on fabrics from different cultures and compare them with the designs on the children's own clothes.
- Take turns to pick up beads from a pile to continue a simple repeat pattern or play picture dominoes.

Language and literacy

- Introduce new vocabulary, such as 'repeat', 'pattern', 'design' and 'symmetry' as you make the pictures.
- Ask the children to choose their favourite patterns and designs from old wallpaper pattern books and use them to make simple home-made books. Write captions about why the children like the patterns they choose.

Mathematics

- Develop the activities by making repeat patterns using small world equipment, such as plastic people and vehicles.
- Thread beads in different patterns onto lengths of string, then challenge the children to copy them. Can they make up their own repeating pattern?

Knowledge and understanding of the world

- Visit the library and look for books about how clothes are made.
- Explore symmetry using mirrors and simple drawings of shapes. Compare the designs created with those seen in the kaleidoscope.

Creative development

- Use fabric pens to draw patterns on squares of cotton sheeting. Use these to make a group patchwork quilt.
- Make a simple tabard style dress for a doll from scraps of tie-dyed fabric by cutting out two identical shapes from a template and sewing them together down the sides.
- Use the sponge print rollers to create designs on paper.

THEMES ON DISPLAY for early years

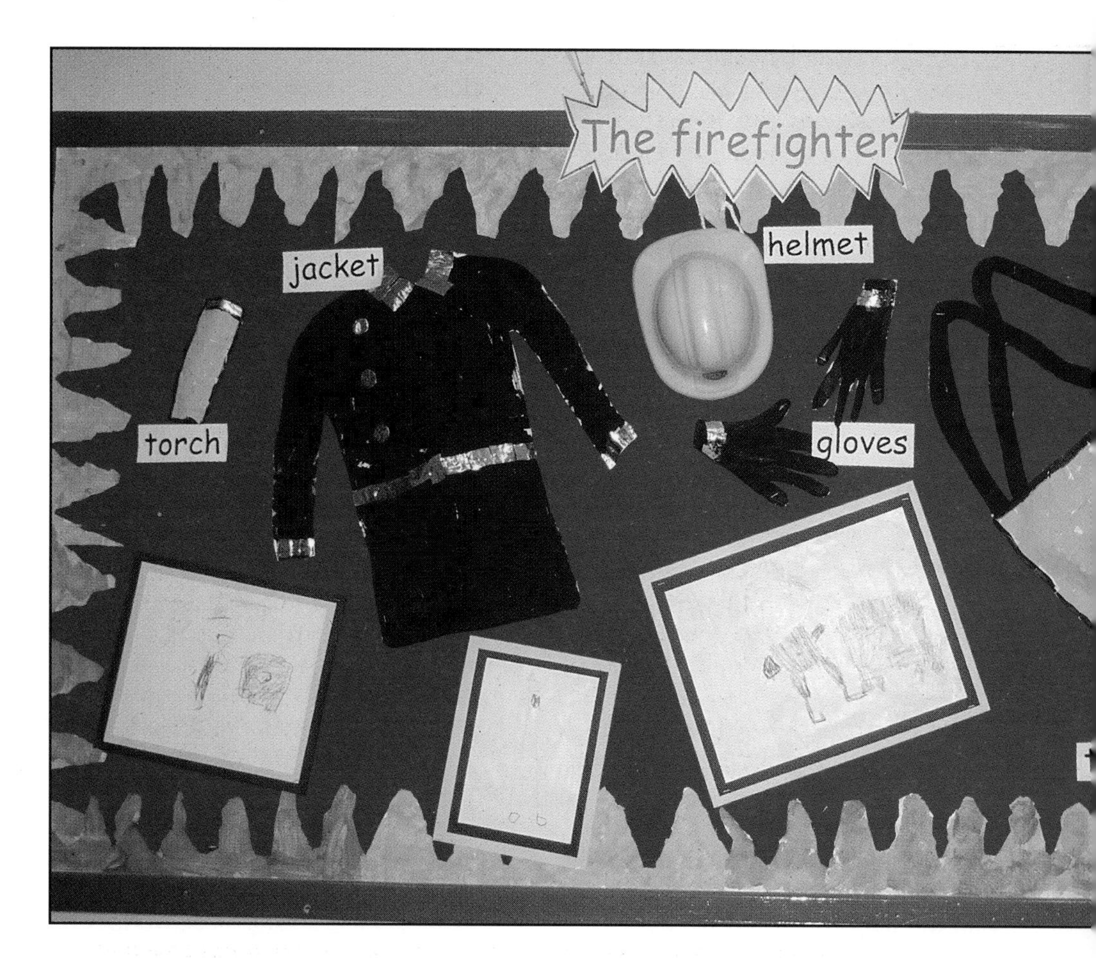

Special clothes – the firefighter

Learning objective: to understand that people wear uniforms either to protect their bodies or as a means of identity.

What you need

Red; black; yellow and white paper; red and yellow paint; sponges; glue; card; collage scraps; firefighter dressing-up clothes; torch; helmet; gloves; red fabric; books such as *Jobs People Do* (Dorling Kindersley) and *Things People Do* by Anne Civardi and Stephen Cartwright (Usborne); samples of clothing and accessories related to other occupations.

What to do

Look at pictures and talk to the children about the uniform of a firefighter. If possible, invite a firefighter to visit the nursery and talk to the children about the fire service. Let the children draw and paint pictures of firefighters and double mount these on black then yellow paper.

Cover the display board in red paper. Cut flickering flame shapes along one edge of wide strips of paper and invite the children to sponge paint them using red and yellow paint mixed together. Arrange the flames around the edge of the display. Alternatively create a flame effect with thick red, orange and yellow paint sponged along strips of white paper.

Draw around a child's jacket, trousers and gloves and cut out the shapes. Let the children collage them to make a firefighter's black jacket and gloves and yellow trousers. Attach to the display board, creating a three-dimensional

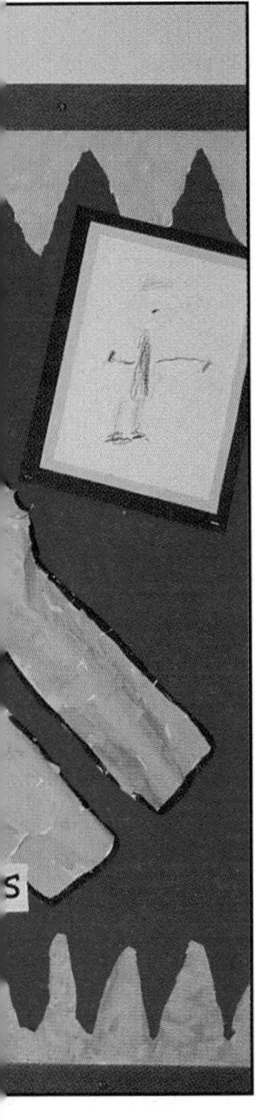

effect by adding scrunched up newspaper behind the shapes, then stapling all the way around. Add black sugar paper braces to the trousers. Cut out a torch shape and collage in the same way. Hang a child-size firefighter's hat onto the display using a cup hook.

Arrange the mounted children's drawings and paintings around the clothes. Make labels and glue these to the backing paper alongside the appropriate clothing. Add a caption for the display in large letters.

Drape a display table with red fabric and arrange the role-play uniforms and hats on top. Include items used by other people who wear uniforms, such as a breathing mask and stethoscope. Display books about uniforms and occupations. Make a caption for the back of the table asking 'Who needs these?'.

An exciting alternative is to display fire-fighting dressing-up clothes on a shop window dummy and stand it alongside the table. Shops may be willing to loan such items for a short period.

Sit with a group of children near the display and let them take turns to choose an item from the table. Discuss the item and what it is used for. Look at the books together. Once the children are familiar with each item let them explore freely by themselves.

Talk about

- Why do people need uniforms to protect them? Talk about the work of a firefighter and the dangers of fire.
- Talk about protective shoes, goggles, reflective strips and gloves. Who wears them? Why? Look at the reflective items worn by a runner or cyclist. How do these protect them?
- Talk about how uniforms are sometimes worn to show that people belong to a special group, such as Brownies or Cubs. Do any of the children's brothers or sisters wear such uniforms?

Home links

- Invite parents who wear special clothing when working to come and visit the children wearing their uniforms and to talk about each item.
- Ask parents whether any of their older children belong to organizations such as Cubs or Guides and invite them to bring in samples of the uniforms.

Using the display

Personal, social and emotional development

- Talk about the need for personal safety and hygiene. Why do the children need to wear aprons, gloves, boots and cycle helmets?
- Discuss the role of the refuse collector in keeping the environment tidy. Have some aprons, gloves, brushes, dustpans and bins available indoors and outdoors and encourage the children to tidy their surroundings.

Knowledge and understanding of the world

- Invite people from different services such as police, fire and ambulance to come and talk about their jobs and the clothes that they wear. Arrange for the children to visit local fire stations on open days.

Physical development

- Dramatize the work of a firefighter, including putting on the uniform, pulling on heavy boots, climbing ladders and holding heavy hose pipes.

Creative development

- Include uniforms in role-play to encourage the children to make up their own stories and develop their imaginative ideas.
- Design a special nursery uniform to wear for the session. Make badges from card and safety pins, hats from card and recycled materials and tabards from scrap fabric.

Baby clothes

Learning objective: to become aware of the passing of time and of the changes associated with growing up.

What you need

Baby gift wrapping paper; paint; collage materials; posters and leaflets from a baby clinic; baby clothes; baby sheet or blanket; baby items such as rattle, bottle and disposable nappy; catalogues of baby clothes; large doll; doll's pram or buggy; basket of baby clothes; baby books such as *The Baby's Catalogue* by Janet and Allan Ahlberg (Puffin) and *You'll Soon Grow Into Them, Titch* by Pat Hutchins (Red Fox).

What to do

Let the children handle the baby clothes and other baby items. Talk about what they feel and look like, then invite the children to make their own collages of baby clothes using tissue paper scraps and other materials to represent the clothes. Mount the finished pictures on bright paper ready for display.

Back the display board in baby gift wrapping paper. Make a border from baby foot and hand prints, cut out and glued to white paper strips. If you have no babies in your nursery ask a parent to bring a baby in, make a sample hand and foot print and then photocopy them. Attach the baby clothes to the display with staples. Make a large caption for the display in the centre of the board and add the children's pictures in the spaces.

Cover a table below the display board with a sheet or blanket and arrange the baby items on it. Place the basket of clothes on the table and sit the doll next to the basket.

Arrange books and catalogues attractively among the baby clothes. Leave the buggy at the front of the table so that the children can dress the doll and take it for a walk.

Talk about

- Introduce the appropriate names for all the items on display. Ask the children if they have younger brothers or sisters and the sort of clothes they wear.

Using the display

Language and literacy

- Make a big book of pictures of baby clothes using children's drawings, photographs and catalogue pictures. Ask the children to suggest captions for you to write alongside the pictures. Encourage those who can to copy the writing. Hang the book in front of the display.
- Read *You'll Soon Grow Into Them, Titch* by Pat Hutchins. Encourage the children to talk about their own place in the family.

Mathematics

- Count the items of clothing for the doll.
- Talk about the order in which the doll's clothes should be put on, using appropriate language.
- Discuss the story about Titch using words such as 'taller' and 'smallest' as you compare Titch with his brother and sister.

Knowledge and understanding of the world

- Talk about baby animals and how they grow. Human babies wear clothes to keep warm. How do animal babies keep warm?

Creative development

- Play a 'feely' game with the items on the table. Cut a hole in the side of a cardboard box with a lid. Put one of the items inside the box and let the children take turns to put a hand inside and feel it. Can they guess what it is? Ask them to talk about what it feels like. Let the children take turns to choose an item to put in the box while the others close their eyes.

Physical development

- Encourage the children to dress the doll in the clothes on the table.

- Pass the clothes around and talk about what they feel like. Are they hard or soft, rough or smooth?
- Talk about the clothes that the children are wearing. How are they different to the baby clothes?

Home links

- Invite a parent with a baby to visit your group and let the children observe the baby's clothes and other special items.
- Ask parents to bring in some baby clothes to show to the children. Do they still have any of the clothes that their nursery child used to wear so that they can bring in examples of baby and pre-school clothes for comparison?

Match the pairs

Learning objective: to match items in pairs.

What you need

Contrasting pairs of shoes of different sizes and functions such as slippers, boots and sandals; shoe cleaning materials; shoe catalogues; cereal box; shoe shop items such as a foot measure and tape; PE bench; table or draw storage unit; books such as *Alfie's Feet* by Shirley Hughes (Red Fox) and 'The Elves and the Shoemaker' (traditional); card; pictures related to nursery rhymes; sticky backed plastic; scissors; black felt-tipped pen.

What to do

Choose some shoe related rhymes such as 'There was an old woman who lived in a shoe', 'One, two buckle my shoe' and 'Cobbler, cobbler, mend my shoe' and make rhyme cards. Find or draw appropriate pictures and write out the rhyme clearly. Cover the cards with sticky backed plastic. Store in a cereal box covered with pictures of shoes. Arrange the bench against a wall and set out the shoes, in pairs, along it grouping them by type. Put the shoe cleaning materials in a shallow tray and display the rhymes in the cereal box. Arrange the catalogues and the foot measure at one end of the bench. Stand the books in spaces at the back of the table or draw storage unit. Let the children explore the area looking at the different types of shoes found and role playing shoe shops.

Choose a child to pick up a pair of shoes and talk about when these might be worn. Look at the shoe cleaning materials and let the children watch as you clean a shoe. (It is better not to let children use shoe polish themselves.) Does it look different after cleaning? What makes shoes dirty? Choose one of the stories to read and say the rhymes, using the cards.

Talk about

- What do we mean by a pair of shoes? What else comes in pairs?
- Why do we need to measure our feet? Talk about visits to shoe shops and having feet measured.

Home links

- Let the children take home the 'Shoe pairs' photocopiable sheet on page 76 and ask parents to help them.
- Ask parents to involve children in the shoe cleaning process.

Further display table ideas

- Display unusual footwear such as flippers, clogs, safety shoes. Discuss who might wear these.
- Have a boot-printing table with a shallow tray of thick paint, a small wellington boot and a large supply of paper.

Food

The display ideas in this chapter provide opportunities for the children to find out about all kinds of foods, with the chance to sample a few!

Healthy foods

Learning objective: to recognize different fruits and vegetables by their colour, shape, smell and taste.

What you need

Green backing paper; white sugar paper; yellow and white card; paint in various colours; paintbrushes; sponges; glue; colouring materials; small shallow plastic trays; fruit and vegetables; magazines; knife (adult use); two tables; cardboard boxes; fruit bowls; mushroom baskets; plastic fruit and vegetables; pictures and posters; plastic grass; books such as *Oliver's Fruit Salad* and *Oliver's Vegetables* by Vivian French (Hodder).

What to do

Invite the children to investigate the fruit and vegetables. Talk about the colours and textures, and smell the food. Cut some fruit in half and notice the pips. Ask the children to make drawings, paintings and prints of the food.

Pour thick red, yellow and orange paint into three shallow trays. Let the children use the halved fruit and vegetables to print a mixed colour border onto strips of white sugar paper.

Cover the display board with green paper and attach the border. Cut a large apple shape from white sugar paper and sponge paint it yellowy green, then glue to the board. Paint a brown stalk.

Cut pictures of fruit and vegetables from the magazines, then stick these onto the surface of the apple. Mount the children's drawings, paintings and prints on yellow card and attach around the apple. Write name labels or make them on the computer and attach to the relevant pictures. Make a large title saying 'Fruit and vegetables'.

Place cardboard boxes on the tables to create display surfaces of different heights. Cover the tables with plastic grass and arrange the books towards the back. Arrange bowls of fruit on one half and baskets of vegetables on the other. Fasten posters and pictures along the front of the tables. Make two labels 'fruit' and 'vegetables' from folded card to stand on each half of the table.

Talk about

- Work with a small group of children. Ask a child to choose a fruit from the table. Pass it around and talk about how it feels and smells. Cut the fruit into small pieces and let the children taste a sample. What does it taste like? Repeat with another fruit then talk about the differences between the two.
- Work with a large group and see how many vegetables the children can name. Pass them around and talk about how they feel and smell.

Home links

- Ask parents to supply items for the display tables.

Where do we buy our food?

Learning objective: to explore colour, texture, shape and form in two and three dimensions.

What you need

Green and white backing paper; large sheets of card; white A4 paper; paint in various colours; paintbrushes; magazines containing pictures of food; scissors; cardboard boxes; brown paper; bright fabric; real and plastic food items; wool.

What to do

Begin by talking about shops with the children. Where do they go shopping? Do they buy all their food from one big shop or do they visit lots of little shops?

Look at pictures of food in magazines. Tell the children that you are going to make a display which shows a row of shops. Decide together which shops to include, and the food that will be 'on sale' in them. Provide A4 paper, painting and colouring materials and let the children draw and paint pictures of food to display in the shops.

Use blue paint to sponge print a sky on white backing paper. Mount this on the top half of the display board and attach green backing paper to the bottom half to represent grass. Using large sheets of card, create a row of three or four shops. Work with the children to use different painting techniques, collage, and the children's own drawings and paintings to make the shops. Paint the name of each shop on white card and staple in place.

Around the shops, add examples of some of the food on display. Make labels and link them to the matching item in the relevant shop with wool.

Place a table in front of the display and cover with bright fabric. Cover cardboard boxes with brown paper, leaving the tops open. Paint on doors and windows and add labels with the name of each shop. Arrange real and

Using the display

Personal, social and emotional development

● Take turns to play a shopping lotto game with a card to represent each shop, divided into six squares. Make smaller cards to fit the squares depicting items sold in the shops.

Language and literacy

● Introduce appropriate shop names, such as 'baker', 'greengrocer', 'butcher' and 'supermarket'. Write the words on cards and match them to the names on the cardboard boxes.

● Make a market stall with some salt dough fruit and vegetables and enjoy role-play.

Mathematics

● Sing 'Five currant buns in the baker's shop', from *This Little Puffin* edited by Elizabeth Matterson (Puffin), choosing children to represent buns or using salt dough buns.

● Count the shops on the display and on the table. Are there the same number? Count the items on the table. Put them into the correct boxes. How many in each box? Which box has the most?

Knowledge and understanding of the world

● Visit the nearest shopping centre and take some photographs of the shops. Put them in a home-made book and write appropriate captions. Hang alongside the display.

● Buy a fish and some prawns from the fishmongers and encourage the children to make observational drawings of your purchases.

Physical development

● Visit a local street and make a list of the shops. Make a model of it using recycled materials.

● Talk about the work of a shopkeeper and dramatize the movements involved, such as stacking shelves, chopping meat, kneading dough and driving delivery vehicles.

Creative development

● Develop a role-play baker's shop by baking salt dough bread and cakes.

● Create model food, such as links of sausages and bunches of grapes, from papier mâché and string.

plastic food items in front of the boxes. Add the phrase 'Can you put us in the right shop?' to the front of the table.

Talk about

● Point to each 'shop' on the display and read the name of the shop to the children. What would the children expect to buy in each one? Talk about visits that the children make to shops. Which shops do they like best? What can they buy there?

● Ask a child to choose an item from the table and to talk about it. Which shop would sell the item? Find the cardboard box 'shop' and put the item in the box.

Home links

● Do any parents, friends or relatives work in shops? Invite them to come and talk to the children about the things they sell.

● Ask parents to play the game with the children, sorting the objects on the table into the correct boxes.

Interactive display

Foods from other cultures

Learning objective: to raise awareness of foods from other cultures.

What you need
Black backing paper; white paper; drawing and colouring materials; samples of packets and dried foods from different cultures; glue; utensils relating to each culture chosen; books such as *The Kids' Round-the-World Cookbook* by Deri Robins (Kingfisher Books); photographs and pictures of food; food samples; fabric; white card.

What to do
Carry out a survey to find out what the children's favourite foods are. Where do these foods come from? Talk about foods from different cultures. Look at the various packets and dried food and talk about the different places around the world that they originate from. Tell the children that you are going to make a display which shows foods from other cultures, and decide on two countries to focus on.

Cover the display board with black paper. Cut scalloped edges from thick strips of paper to use as a border. Draw outlines of the flags from each country along the strips of paper and let the children use coloured pencils to colour them in. Staple the border to the display board, dividing the two halves down the middle. Colour in sheets of A4 paper to represent each flag, and use these to make large labels for each section. Attach empty and full packets, dried food samples, small boxes, books, pictures and children's drawings and paintings to the relevant sections.

Choose one of the cultures as a focus for the display table. Cover the table with fabric and arrange the appropriate items on top. For example, for a Chinese table, include chopsticks, bowls of rice or prawn crackers, tinned and fresh lychees, books, pictures and photographs. Make a card label saying 'Chinese food'.

Talk about
- Discuss the food items. Have the children tasted any of them? Pass around the bowl of prawn crackers and a bowl of crisps. Which do the children prefer? NB: Check for food allergies before tasting any of the food.

Home links
- Encourage parents to look for countries of origin on packaging with their children and to bring in samples of this packaging.

Using the display

Personal, social and emotional development

● Talk about Italian food and cook some pasta. Try adding things to alter the flavour, such as grated cheese, tomato sauce or chopped herbs.

● Does the timing of the display coincide with a festival, such as Hanukkah, Eid or Easter? Choose traditional foods associated with this festival to sample.

Language and literacy

● Read *Cleversticks* by Bernard Ashley (Picture Lions) about a Chinese boy who impresses his friends at nursery with his ability to use chopsticks.

● Make 'step by step' illustrated recipe cards for the preparation of simple dishes from the countries featured. Laminate and display the cards.

Mathematics

● Follow one of the recipe cards, introducing appropriate mathematical language and skills.

● Compare pasta by shape, size and colour. Measure strips of spaghetti before and after cooking.

Knowledge and understanding of the world

● Find out more about the countries featured using atlases, globes and books.

● Make direct comparisons between raw and cooked food. Weigh raw and cooked pasta or rice, taste raw and cooked celery, handle raw and cooked prawn crackers. Take photographs and use these alongside drawings and written observations to produce a book for the display.

Physical development

● Have fun playing a game which involves picking up foods such as cooked and uncooked pasta, using a range of tools and utensils such as forks, spoons, chopsticks, tweezers and tongs. Time how long it takes to pick up a selection of items and transfer them to a dish.

Creative development

● Use magazine pictures and empty packets to create collages representing food from a particular culture.

● Fill an empty water tray with cooked pasta. Let the children handle the mixture. Compare the texture with that of a bowl of uncooked rice. Which do the children like handling best?

Interactive display

Foods that grow underground

Learning objective: to compare different vegetables and use appropriate mathematical language associated with their growth.

What you need

Copy of *The Enormous Turnip* illustrated by Jan Lewis (*First Favourite Tales* series, Ladybird); neutral backing paper; white border paper; purple and orange fabric scraps; string; paint in various colours; card; cardboard 'comb'; sponge; tissue paper; green plastic carrier bags; toothbrush or nail brush; carrots, onions and beetroot; plastic 'grass'; 'hyacinth' jar; books related to the theme; tinned carrots; dried onions; pickled beetroot; seed packets; soup and sauce mixes; empty frozen vegetable bags.

What to do

Share the story of *The Enormous Turnip* with the children. Talk about how the turnip grew bigger and bigger underground, until it was so huge that the man could not pull it up. Talk about other foods which the man could have grown in his garden which also grow underground. How many can the children name? Tell them that they are going to find out more about foods which grow underground by helping to make a display.

Prepare the background for the display using neutral sugar paper. Sponge print it using different shades of brown and blue paint to represent the soil and the sky. Use a toothbrush or nailbrush to splatter paint on to add texture. Cut a border with a scalloped edge and let the children use potatoes and carrots cut in half to print patterns. Staple the border to the display board.

Cut out the outline of a large beetroot from sugar paper. Invite the children to choose appropriate colours to paint the outline. When it is dry, add texture by gluing on fabric scraps in different shades of purple. Make tiny hair-like 'roots' by dipping string in purple paint and create large bushy 'leaves' from

green carrier bags or scrunched up tissue paper. Use a similar method to make a large 'carrot' from orange fabric.

Create a large 'onion' by swirling thick brown paint over the surface of the card using a cardboard comb. Attach the 'vegetables' to the wall display.

Add underground creatures and stone shapes using collage materials. Make beetles from black plastic refuse sacks, worms from pink wool, and stones from thick fabric.

Make labels and attach them in the relevant places. Add labels using appropriate mathematical language to indicate position, size and shape, such as 'tall' onion, 'long' root, 'straight' stalk.

Cover the table with the plastic grass cover. Display samples of carrots, onions and beetroot for the children to handle. Add an onion in a jar with visible roots, samples of tinned carrots, dried onions, pickled beetroot, packets of onion, carrot and beetroot seeds, soup and sauce mixes, empty frozen vegetable bags, gardening books and recipe books.

Talk about

- Emphasize the use of mathematical language by talking about the parts of the vegetables which are 'over' and 'under' the ground, making comparisons between the 'long, thin' carrot and the 'round, fat' beetroot, and introducing words like 'tall', 'straight', 'wide' and 'narrow'.
- Talk about how vegetables are preserved in tins and by pickling, bottling, drying and freezing. Look at examples of each.

Home links

- Ask parents or relatives who grow their own vegetables to bring in samples, preferably straight from the ground before washing.
- Invite parents to come and help with soup making.

Using the display

Personal, social and emotional development

- Try growing beetroot and carrot tops, letting the children water them daily. Grow potatoes in a bucket.
- Encourage turn taking, sharing and concentration skills by preparing vegetables for home-made soup.

Language and literacy

- Read the traditional stories 'The Enormous Turnip' and 'Jack and the Beanstalk'. Talk about the children's experiences of growing vegetables.
- Make observational drawings of vegetables. Label the roots, stalk and leaves.

Mathematics

- Sort a box of mixed vegetables into different types. Count how many of each kind there are.
- Colour the 'The enormous turnip' photocopiable sheet on page 77. Discuss the mathematical vocabulary.

Knowledge and understanding of the world

- Experiment with the conditions needed for growth. Keep some seeds in the dark and others in the light, water some and leave others dry. Observe and discuss your findings.

Physical development

- Dramatize the story of 'The Enormous Turnip' emphasizing the movements needed to pull up the huge turnip.
- Use a range of tools, such as small spades, spoons, trowels and forks, to sow seeds in plant pots and decide which are the best for the task.

Creative development

- Move to appropriate music and pretend to be roots stretching down into the soil and stalks growing up to the light. Introduce percussion to represent the rain and wind and move in response to the sounds.
- Create models of vegetables using clay, salt dough or papier mâché. Follow the salt dough recipe on the photocopiable sheet on page 78.

Dairy foods

Learning objective: to look for similarities, differences and changes in milk and other dairy products.

What you need
Green and blue backing paper; green fabric scraps; green, white, black and yellow paint; white card; white fabric; dairy items such as milk cartons, cheese boxes and yoghurt pots; plastic cows; toy milk lorries; recycled materials including cardboard tubes; books such as *Am I Fit and Healthy? Learning About Diet and Exercise* (Wayland); milk shake cartons and mixes.

What to do
Ask the children to help you create a country scene on the display board. Begin by backing the board using green and blue backing paper to represent grass and sky. Staple on scraps of green fabric to make tufts of grass. Make clouds from scrunched up white tissue paper stuck to white card and a yellow 'sun' from a painted card circle. Invite the children to work in small groups to paint cows using black and white paint. Cut these out and stick them to the display board. Cut out milk bottle shapes from white card and stick them around the edge of the display to form a border. Use recycled materials, such as cardboard tubes, to create model cows and paint them black and white.

Cover the table with white fabric and display the dairy items and children's models on the table among the books.

Talk about
- Discuss the stages in the journey of milk from the cow to the fridges in the children's homes. Use books and posters for reference.
- Look at the products made from milk. How many can the children name?
- Include plastic milk cartons in water play.

Home links
- Ask parents to collect packaging from dairy products for the display.
- Invite parents to help with making milk shakes.

Using the display

Personal, social and emotional development

- Make skeleton collages from pieces of white paper and talk about how milk helps to build strong bones and teeth.
- Make fruity milk shakes by blending chopped fruit with milk. Encourage the children to choose which fruit to include and to decide which flavours they prefer.

Language and literacy

- Write to the National Dairy Council at 5–7 John Princes Street, London W1M 0AP for posters and leaflets. Visit their website at www.milk.co.uk
- Read poems and sing songs about farm animals, such as 'Old MacDonald'.
- Include empty milk cartons, a white coat and a plastic milk crate in the home area to encourage children to develop role-play involving the delivery of milk.

Mathematics

- Encourage children to recognize and use numbers to ten by playing games with empty milk cartons. Draw around a carton base on a strip of card and repeat to form a row of ten squares. Number the squares. Put a carton on each square and count them.
- Introduce simple addition and subtraction by adding and taking away cartons along the strip.

Knowledge and understanding of the world

- Look at models and pictures of cows. Compare different breeds. Paint and draw cows and make models referring to these pictures and models.
- Arrange a visit to a dairy farm or local dairy. If this is impossible ask someone who delivers milk to visit and talk about what the job involves.

Physical development

- Play a milk delivery game using trolleys or wheelbarrows to transport crates of empty cartons or plastic bottles. Draw chalk roads for the 'milk vehicles' to follow. Draw in houses and make signs to hang over the backs of chairs to indicate how many cartons are required at each 'house'.

Creative development

- Use empty milk and yoghurt cartons to create model vehicles with cheese box 'wheels'.
- Make a model farm layout on a board, using papier mâché to create meadows. Paint grass and hedges. Make farm buildings from boxes. Develop play to include the cows being milked in the milking parlour.

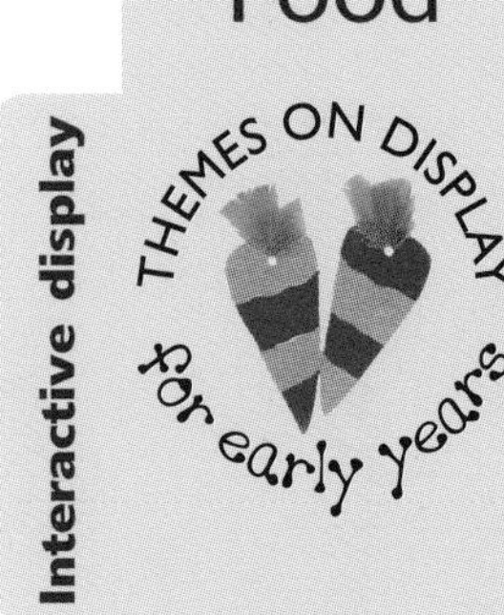

Party time

Learning objective: to respond to special occasions in a variety of creative ways.

What you need

Bright backing paper; children's birthday wrapping paper; party plates; party hats; drinking straws; silver foil; party tablecloths; birthday cards; painting and drawing materials; A4 paper; collage materials; small circular table; small chairs; dolls; plastic plates and cups; salt dough food; small parcels; balloons; magazines containing pictures of food; scissors; glue; books about parties; items associated with parties such as jars of sweets and party crackers.

What to do

Gather the children together and talk about birthdays. Who has had a birthday recently? Did they have a birthday party? What sort of food did they eat at their party? Tell the children that they are going to make their own party scene including lots of different types of party food.

Cover the display board with bright backing paper. Create a border from strips of wrapping paper and tie balloons to the corners of the board. Attach a party tablecloth to the centre of the board or print your own. Provide painting materials and ask the children to paint pictures of their faces. Give them each a paper hat and let them stick a hat onto their painting. Arrange the completed faces around the tablecloth.

Prepare some party food by sticking items cut from magazines or made from collage materials to the party plates. Attach the plates the cloth on the display. Make 'glasses' from foil circles. Arrange birthday cards, both commercially produced and home-made, small parcels, drawings, paintings and posters in the spaces around the tablecloth. Label the items. Arrange the party books and objects attractively in front of the display.

Cover a display table with a party tablecloth and set out a doll's party using plates and cups from the home area and salt dough 'food'. Sit the dolls on chairs around the table and put party hats on them.

Talk about

● Talk about the different foods that the children made pictures of or cut from

magazines. What are the children's favourite party foods?

- Talk about the wall display. How many children are at the party? What are they eating? What might be in the parcels?
- Discuss the doll's party. Are there enough hats, plates and cups for the dolls?

Home links

- Ask parents to save old birthday cards and wrapping paper.
- Invite grandparents to come in to your setting and talk about parties when they were young.
- Ask parents to bring in photographs of past and recent parties the children have attended.

Using the display

Personal, social and emotional development

- Have a party with special food to celebrate a cultural festival or event.
- Have a simple 'grandparents party' and ask each child to invite an elderly relative or neighbour. (Parents may need to suggest someone.)

Language and literacy

- Design birthday cards or cards for other special occasions.
- Write and illustrate menus for dolls' and grandparents' parties.
- Make special name cards for the children's snack time.

Mathematics

- Extend 'party' play into the home area. Set the table for a party, making sure that there is a plate, cup, hat, napkin and place name for each child or doll. Count the items and the party 'guests' to check.
- Make napkins by colouring squares of paper. Talk about the shape and fold the napkins into triangles.
- Set the table for three different-sized teddy bears using large, small and medium-sized items and introducing appropriate language into the play.

Knowledge and understanding of the world

- Make a jelly or some fruit ice lollies for a party. Discuss changes between solid and liquid. Introduce and explain the meaning of appropriate words, such as 'set', 'pour', 'freeze' or 'melt'.
- Use parents' photographs of past and recent children's parties to make a display of 'Parties past and present'. Talk about the changes in the children as they grow.

Physical development

- Enjoy playing party games which involve moving freely and responding to instructions, such as 'musical statues' and 'musical chairs'.
- Have a 'dance' and encourage the children to respond imaginatively to the music.

Creative development

- Play a guessing game using toys which make a noise such as a baby's rattle or a toy with bells. Place them in identical boxes and wrap them up as 'presents'. Challenge the children to identify them by sound alone.
- Make party place mats using brightly coloured and textured paper.

Picnic time

Learning objective: to handle appropriate tools and materials safely and with increasing control.

What you need
Recycled materials including shoe boxes with lids; glue; spreaders; scissors; masking tape; ribbon; string; wool; paint; wrapping paper; salt dough; paper napkins; paper cups; plastic grass cover; newspaper; samples of lunch boxes.

What to do
Prepare for this display by making some salt dough food items suitable for a lunch box and painting them. Explore the size, shape and colour of the sample lunch boxes and explain to the children that they are going to make their own lunch boxes from a selection of recycled materials.

Let each child choose a box and either cover it with wrapping paper or paint it. Do the same with the lid, ensuring that it matches the box. Talk to the children about how they could make handles and let them experiment with ribbon, string, wool and tape. Line the box with a napkin and fill with salt dough food and a paper cup.

Screw up pieces of newspaper and stick them to the table with tape. Cover the table with the plastic grass so that the newspaper creates a 'hilly' effect. Alternatively, spread a picnic cloth on the table. Arrange the lunch boxes open on the table.

Talk about
● Let the children take turns to take their lunch box off the table and talk to the others about how they made it and the 'food' it contains.
● Talk about the size and shape of the lunch boxes. Are they easy to carry? Will the food fit in without being squashed?

Home links
● Ask parents to send in empty lunch boxes as samples.
● Invite them to join a pretend teddy bear's picnic, either outdoors or on the carpet.

Further display table ideas
● Draw around some plastic food shapes and cut the outlines out. Stick them on the lids of the lunch boxes and put the appropriate item of food inside. Have a guessing table and try to guess what is in each box by the shape on the lid. For older children try writing the initial letter sound of the object instead of drawing round the outline.
● Make a 'fast food' table. Let the children make their own 'branded' cups, burger boxes and napkins from a variety of materials. Add plastic straws, aprons, caps and other relevant items, then use the items for role-play.

Toys

All children love toys, and the displays in this chapter provide many opportunities for hands-on experience with a wide variety, plus opportunities for children to create their own toys.

Toys past and present

Learning objective: to develop an awareness of past and present.

What you need

Bright backing paper; samples of toys from the past suitable to attach to the wall (perhaps in a see through bag); pictures of toys from the past; scissors; glue; beige sugar paper; sponges; brown paint; fabric; A3 size coloured sugar paper; children's favourite toys; books about new and old toys; thick paper; felt-tipped pen; wool and needle; photographs of the children with their toys; drawing materials; white paper.

What to do

Mount the backing paper on the board. Arrange the old toys in the centre of the space and attach them securely. If they may be damaged, put them in clear bags and attach the bags instead. Include the pictures of old toys. Add a label for each item.

Make the display look like an old shop window by attaching a border of beige sugar paper sponge painted brown. Add thin strips of sponge painted sugar paper to represent window frames across the space.

Make home-made books from individual sheets of A3 size sugar paper. Let the children draw pictures of their favourite toys and stick them onto the sheets. Add captions together and stick in photographs where appropriate. Sew the pages together with thick wool (an adult should do this).

Place a table in front of the board and cover with fabric. Ask the children to arrange their favourite toys on the fabric. Display the books, including home-made examples, among the toys.

Talk about

- Discuss the differences between the children's toys and those from the past.
- Encourage the children to talk about their favourite toys. Why do they like them?

Home links

- Encourage parents or grandparents to show their toys to the children and to talk about the things they used to do and the games they used to play when they were children.

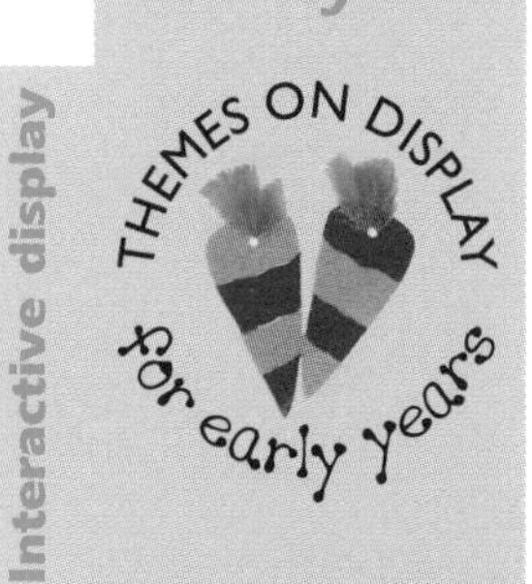

Favourite toys

Learning objective: to use the senses of touch, sight and hearing to recognize and identify objects.

What you need

Yellow backing paper; border roll depicting toys; 'toy' stencils or sponge shapes; paint in various colours; black and blue paper; coloured card; sticky tape; toy catalogues; glue; scissors; cardboard boxes; parcels containing toys covered in children's wrapping paper; books such as *Baby's Toys* by Fiona Watt (Usborne) and *Kipper's Toybox* by Mick Inkpen (Hodder); bright fabric.

What to do

Begin by talking about the children's favourite toys. Have they had the toys since they were babies, or are they new toys? Share the catalogues with the children and invite them to point out favourite toys, then help them to cut out the pictures. Choose a low display board and cover it with yellow paper. Let the children help to print toy shapes all over using the sponges or stencils, then add the border. Add the title 'Our favourite toys' and attach the children's cut out pictures.

Find fairly large pictures of recognizable toys in the catalogues and cut them out carefully. Draw around the

pictures on black paper and cut them out to form the outline of different toys. Stick these black outlines onto separate sheets of blue A4 sugar paper. Stick the toy pictures to the display board first, and cover them with the corresponding A4 picture, secured at the top with tape to make a lift-the-flap picture. The children can try to guess what the toys are before lifting the flaps to see if they guessed correctly.

Place different sized cardboard boxes on the floor in front of the display and cover with the fabric. Arrange the wrapped parcels, books and catalogues on the boxes and mount a sign saying 'Guess what's inside'.

Talk about

- Discuss the outline pictures in turn. Can the children guess what might be underneath? Does the toy have wheels? What shape is it?
- Ask the children to choose a parcel and describe how it feels to the others.
- Talk about toys suitable for babies. What safety factors are important when choosing these?

Home links

- Ask parents to let their child bring in a favourite toy on a particular day for a 'show and tell' session.
- Invite parents to help create the display or play the guessing activities.

Using the display

Personal, social and emotional development

- Ask children to bring in special toys. Talk about the importance of handling them with care and respecting the property of others.

Language and literacy

- Write labels for the toys in the display and talk about the initial letter sounds of the words. Can the children identify a toy by saying the sound of the first letter? Play 'I spy' with the toys on the display.

Mathematics

- Sit a group of dolls on the floor beside the display and share out the parcels. Is there a parcel for each doll? How many more are needed? Are there too many? How many should be taken away?
- Cut out toy pictures and sort them into different categories, such as toys with wheels, baby toys, dolls or games. Make a book of the results.

Knowledge and understanding of the world

- Wrap up some parcels for the display, games and role-play activities using a range of materials and resources for joining. Let the children experiment to decide which are the most appropriate materials to use.
- Play 'pass the parcel' with some dolls using a battery operated tape recorder. Let the children operate the controls themselves and unwrap the parcel for each doll.

Physical development

- Dramatize the movement of toys suggested by the children, for example move like cars and trains, roll and bounce like balls or spring like a jack-in-the-box.
- Wrap some miniature parcels for a 'small world' party using tiny items, small pieces of paper and wool.

Creative development

- Try printing patterns and lines with small plastic toys, such as car tracks, dolls' foot prints, rolling marbles and dinosaur tracks.
- Try to identify toys by their sounds such as a drum, trumpet, rattle, squeaking mouse and growling teddy.

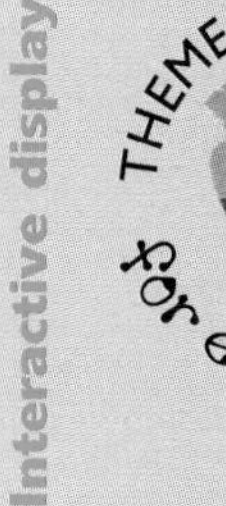

Mechanical toys

Learning objective: to question how things work and use technology appropriately to support learning.

What you need

Black backing paper; silver foil; silver paper; glitter; sequin waste strips; white paint; sponge; recycled materials including cardboard tubes and boxes; battery-operated and mechanical toys; pictures of mechanical robots in toy catalogues.

What to do

In small groups let the children investigate the toys. What is special about them? Tell the children that all these toys move in some way, and demonstrate each one. Respond appropriately to the children's questions about how the toys move. Challenge the children to suggest other toys which move. If necessary, suggest mechanical robots and show them pictures in catalogues. Tell the children that you are going to make a mechanical robot display.

Cover the display board with black paper and sponge print all over the surface with white paint. Create a border from strips of silver paper. Invite the children to search through the recycled materials for suitable items to create a robot. Help them to choose suitable sized boxes to make the head, body and feet, and tubes for the arms and legs. Cover the items with foil and assemble the robot on the board using glue, large drawing pins and sticky tape. Add features using other recycled materials covered in foil.

Cover the surface of a small table with foil and arrange a selection of mechanical and battery-operated toys, with buttons to press and handles to turn, for the children to explore.

Talk about

- How might the robot talk? Imitate the sound of an electronic voice. Could a robot sing, laugh or cry? What can the children do with their voices?
- Make a list of words to describe animal movements such as 'jump', 'gallop' and 'slither'. Compare with words describing a robot's movements, such as 'stiff', 'slow' and 'heavy'.

Home links

- Ask parents to help children to make a list of tasks around the house that they would like a robot to do for them. Use the lists as the basis for a discussion about what children imagine a robot can do.
- Ask parents to supply suitable recycled materials and to come and help at a 'robot making' session.

Using the display

Personal, social and emotional development

- Talk about the idea of robots helping to do housework as a starting point for a discussion about how children can help at home and share tasks with others.
- Robots cannot express their feelings. They never smile, laugh or cry. Ask the children to paint pictures of laughing, smiling and sad faces and talk about what might be causing their happiness or sadness.

Language and literacy

- Introduce robots into role-play in the home corner using simple silver tabards. Explain how robots move and how they can with help simple household tasks.
- Make a large robot outline on the wall. Ask the children to draw pictures of things starting with the letter 'r' and write the word alongside. Include an appropriate caption, such as 'Can you read the words inside Rob the robot?'.

Mathematics

- Provide foil scraps in different shapes, including rectangles, circles, triangles and squares. Let the children make individual collage pictures of robots using the shapes. How many shapes can they identify?
- Ask the children to sort silver collage materials into separate containers according to their shape. Count how many items are in each container. Grade them for size.

Knowledge and understanding of the world

- Explore the toys on the table and discuss how they work.
- Make a moving toy using a cotton reel with an elastic band threaded through and a matchstick attached to each end. Wind up the elastic by turning one of the matchsticks tightly, release and watch your robot slowly move across the floor.

Physical development

- Create an assault course with tunnels to crawl through, slopes to slide down and beams to balance on. Try moving along the course like a robot with stiff arms and legs. Contrast this by pretending to be snakes and slithering along the course. Finally ask the children to be themselves as they complete the course. Talk about the difficult areas encountered by the robot and snake. Were they the same?

Creative development

- Explore sounds made when the same materials are knocked together, such as metal lids against spoons or wooden sticks on blocks. Discuss what a robot would be made of and choose appropriate metallic sounds to represent a robot moving. Use these to accompany children as they move stiffly about the room.
- Listen to extracts from *Coppelia* by Delibes and *The Nutcracker Suite* by Tchaikovsky. Talk about the mechanical doll and toys in the story and move imaginatively to the music.

Kites

Learning objective: to use appropriate language to describe shape and to recognize and recreate mathematical patterns.

What you need

Blue backing paper; brightly coloured tissue paper; string; wool; white paper; light blue paint; sponge; dowelling; glue; pieces of plastic carrier bags; fabric; glitter; sequins; tinsel; books such as *The Windy Day* by Mick Manning and Brita Granström (Franklin Watts) and *The Wind Blew* by Pat Hutchins (Red Fox); kites.

What to do

Involve the children in some kite-making sessions using a wide range of materials. Begin by looking closely at a kite and identifying the main parts, such as the frame, tail, string and string holder. Talk about how it is made, how the frame needs to be strong to support the kite yet light so that the kite can fly in the sky. Discuss the design and decide which patterns the children prefer. Try out the kites on a windy day, and keep some for the display.

Cover the board in blue paper and sponge print it with light blue paint to create a 'windy' sky. Add white clouds using fabric, paper or tissue. Create the title 'Kites' in dark lettering and glue to the cloud. Make a border from twists of coloured tissue wound round string to represent kite tails. Pin some of the children's kites to the display and hang others from the ceiling as mobiles.

Arrange tools and materials suitable for kite-making on the table below so that the children can experiment. Include glue, paint, dowelling, scissors, fabric and plastic, tissue, and smaller items for decoration. It may be necessary to have an empty table nearby for the children to take their chosen resources to assemble the kites.

Using of the display

Personal, social and emotional development

- This display would link well with a mini-topic on Japanese Children's Day on 5 May, when carp-shaped kites are traditionally made and flown on poles in gardens.
- Make and fly kites in pairs, encouraging co-operation and sharing.

Language and literacy

- Make a list of words associated with flying and windy days, such as 'float', 'blow', 'stormy'. Write them inside mobile kite shapes.
- Make up a story about a kite's journey. Where will it land? Which places will it visit?

Mathematics

- Create repeat patterns on the kites with tissue circles, felt tipped-pens or small collage scraps.
- Use appropriate language to describe the position of kites in the sky, for example, 'highest', 'lowest', 'higher', 'above' and 'below'.

Knowledge and understanding of the world

- Explore ideal weather conditions for flying kites. Keep a weather chart daily and record when the kites are able to fly. Will they fly on days without wind or when it is raining? Question why.
- Try making a wind sock from an old pair of tights or a carp-shaped Japanese kite from tissue paper. Make daily observations of wind direction and speed. How can you tell when the wind is blowing harder?

Physical development

- Take the home-made kites outside and try running with them. How fast do the children need to run to make the kite fly?
- Make other models which fly, such as parachutes and paper aeroplanes, using a range of tools and materials to increase manual dexterity.

Creative development

- Pretend to be kites dancing through the air, tossing and turning, twisting and swirling to background music. Hold out long tissue paper streamers to represent the kite tails.
- Make paper kite frames and paint patterns or faces using thick, brightly coloured paint.

Interact enough to give guidance and physical help if necessary, but allow children freedom to experiment. Display relevant stories and books alongside.

Talk about

- Discuss how different things fly. Do they have wings, an engine, rotary blades, propellers? Would the children like to fly? How would the view from the sky differ from that of the ground? Talk about journeys in an aeroplane.
- Talk about windy days. Use vocabulary to describe the action of the wind, such as 'buffet', 'swirl' and 'gust'.

Home links

- Send home instructions for parents to make a basic kite with their children.
- Invite parents to come in to help with kite-making sessions and flying the finished kites.

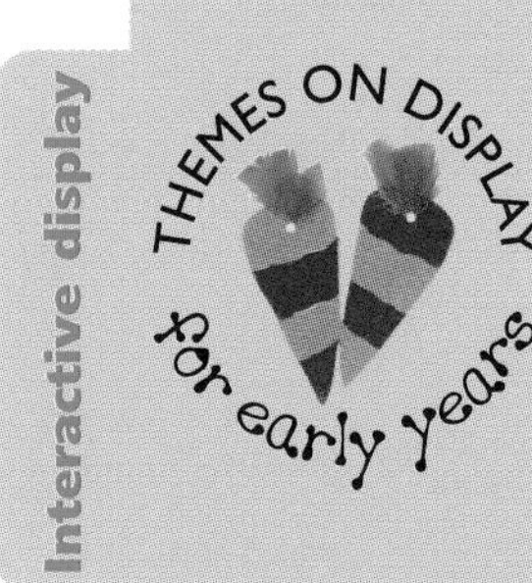

Puppets

Learning objective: to express thoughts and feelings and to make up stories through imaginative play.

What you need

Bright backing paper; textured wallpaper; contrasting fabric; foil; paint in various colours; decorative cord; two cup hooks; cardboard box; recycled collage materials; wool; green crêpe paper; paper bags; card; dowelling; blunt needles; examples of puppets.

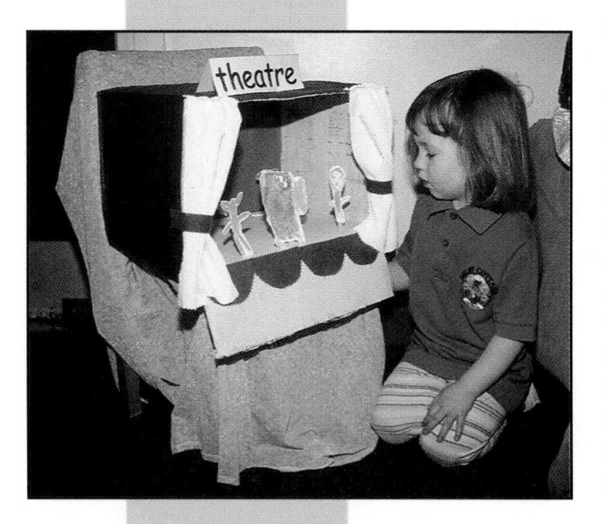

What to do

Talk about puppets with the children. Have any of them ever seen a puppet show? Maybe some of them own puppets. Introduce the examples of puppets and demonstrate how they move. Let different children try moving the puppets, giving directions as necessary. Invite the children to create their own puppets and a puppet theatre, using a variety of materials.

Begin by creating a theatre scene on the display board. Cover the board with bright backing paper. Create an elaborate curtain effect by stapling fabric across the top to form a pelmet. Attach fabric to either side and use cup hooks and decorative cord to tie them back. Use textured wallpaper painted brown to represent the stage and foil circles for footlights. Add the name of the theatre at the top of the display.

Make puppets using a variety of different methods.

Paper bag puppets

Ask the children to draw faces on the bags and put their hands in to make the puppets move.

Pop-up rabbit puppets

Make a cone shape from cardboard or use a paper cup. Cover the cone with green crêpe paper and cut a fringe around the widest edge at the top to represent grass. Leave a hole in the base of the cone. Cut out a rabbit shape from card, paint it and add features. Attach the rabbit to a piece of dowelling and push the dowelling through the bottom of the cone until the rabbit is 'hiding' in the cone. Make the rabbit 'pop up' by pushing up the dowelling. Create different characters using this method.

String puppets
These can vary in complexity from a simple card character hanging from a string to one with moving limbs joined with paper fasteners and with several strings attached to head, arms and legs.

Hand puppets
Cut out two identical hand puppet shapes and encourage the children to try sewing the sides together with large blunt needles and wool.

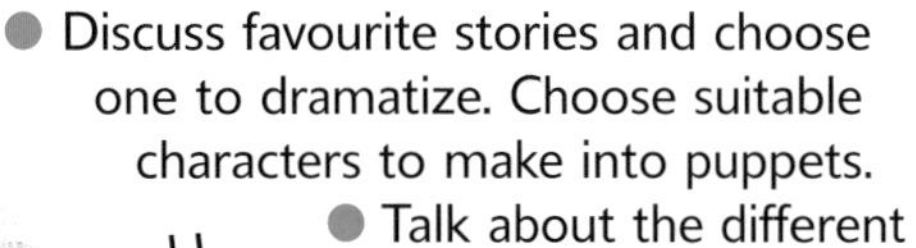

Finger puppets
Cut out two identical character shapes from felt, slightly wider than a finger. Glue them together and attach felt features.

Attach a variety of the children's puppets to the display board theatre and label them.

Make a miniature version of the display using a cardboard box. Paint the inside of the box to represent the stage and add fabric curtains. Cut a slot along the front edge in which to stand the children's stick puppets.

Talk about
- Discuss favourite stories and choose one to dramatize. Choose suitable characters to make into puppets.
- Talk about the different methods of making puppets. Decide on the materials and colours to use.

Home links
- Ask parents to send in examples of puppets for the children to look at.
- Invite parents to watch a puppet performance.

Using the display

Personal, social and emotional development
- Encourage each child to choose a puppet and use it to perform a favourite rhyme in front of the others.
- Make up small scenes using the puppets. Use these to stress the difference between right and wrong.

Language and literacy
- Use a puppet to encourage the children, particularly those who are shy, to talk about their experiences by asking them appropriate questions.

Mathematics
- Use finger puppets to make number rhymes more stimulating. Try 'Five Little Ducks', 'Five Little Speckled Frogs' and 'Ten in the Bed'.
- Give a child a hand puppet to put on and then ask questions inviting the puppet to put an object in a particular position, such as inside the box. Have a puppet of your own to correct any of the child's uncertainties by saying 'Shall my puppet have a try now?'.

Knowledge and understanding of the world
- Arrange a visit to a local theatre or arts centre to watch a puppet show or children's production.
- Use the puppets to encourage the children to talk confidently about their families and past and present events in their lives.

Physical development
- Present the children with a wide range of tools and materials to make a puppet theatre, starting with a cardboard box, so that they can develop increasing manual dexterity as they experiment.

Creative development
- Make a puppet theatre from a large box and let the children decorate it. Use it to put on a performance. Record a story, rhyme or music on a battery-operated tape recorder so that children can operate it themselves. Leave a box of puppets nearby so that the children can use them to do the actions.

Toys which move

Learning objective: to handle appropriate tools and materials safely and with increasing control.

What you need

Bright backing paper; large sponge; recycled materials; black wool pom-poms; black card; pipe-cleaners; string; paper; paint; paper plates; paper fasteners; plastic cup hooks; sticky paper; cotton reel; cover for table or floor; toys which move such as cars, yo-yos and a jack-in-the-box.

What to do

Choose a low display board at the children's height and cover it with bright paper. Cut out a large spider shape from sponge and create a border of black painted spider prints. Create the title 'Make it move'.

Show the children the selection of toys which move and let them investigate the toys. Can they tell you which part of each toy moves? Invite them to talk about other moving toys that they have at home. Now show the children the recycled materials and tell them that they are going to make their own moving toys.

Spider on a string

Make a spider by gluing pipe-cleaner 'legs' to a small black wool pom-pom or cardboard cone shapes. Add sticky paper eyes. Attach a long piece of string or black wool to the spider. Screw a cup hook into the display board and thread the spider over the hook. Attach a cotton reel to the other end of the wool or string to weight it down. Pull the reel up and down to make the spider move.

Moving eyes

Using the photocopiable sheet 'Moving eyes' on page 79 as a template, let each child child make and decorate their own face with 'moving' eyes.

Mobile teddy

Cut out a separate body, head and limbs for a teddy from card and join them together with paper fasteners. Pin the teddy to the display. The limbs can be moved up and down manually by the children.

The mouse ran up the clock

Make a cardboard clock shape and attach a fabric mouse on a string. Thread the string through the clock face so that it hangs down at the back. Pin the clock to the display. Pull the string to make the mouse run up the clock.

Experiment with other toys using similar techniques. Spread the moving toys on the cover on the floor below the display, or a nearby table, so that the children can explore them freely.

Talk about

- Compare how the home-made toys work with the toys on the floor.
- Ask each child to choose a favourite toy from the selection on the floor and to talk about it to the others.

Home links

- Ask parents to help with making the home-made toys for the display.
- Encourage parents to play with the toys on display with their children.

Using the display

Personal, social and emotional development

- Introduce group play with the toys on the floor and encourage sharing and turn-taking.

Mathematics

- Sort and match small world cars according to type and colour. Count how many in each group.
- Put a group of five 'small world' people in a line. Ask the children how many cars will be needed so that each person has a car. Make a garage for each car using cardboard boxes. Introduce addition and subtraction into the play. For example, tell the children that two cars have broken down and must stay in the garages. How many cars are working?

Knowledge and understanding of the world

- Create slopes with planks and roll wheeled toys down them to investigate how the degree of slope affects the speed of the vehicle and the distance travelled.

Physical development

- Pretend to be a jack-in-the-box, curling up tightly and then springing up to the sound of a tambourine.

Creative development

- Make models of other favourite nursery rhymes such as 'Jack and Jill' and 'Little Miss Muffet'. Create a simple pulley to hoist a yoghurt carton 'bucket' out of the well and a dangling spider on a piece of elastic to frighten Miss Muffet.

Make your own games

Learning objective: to work as part of a group, sharing and taking turns.

What you need

Bright fabric; white card; black felt-tipped pen; colouring and painting materials; laminating materials; scissors; plastic bottles; sand; PVA glue; balls; marbles; paper; shoe box; string; dowelling; plywood; children's puppets.

What to do

Cover the table with bright fabric and add a title 'Homemade Games'. Prepare a range of games with the children.

● Skittles – make the skittles out of ten plastic bottles. Pour dry sand into the base to make them stable, spread PVA glue around the top and screw the lids on tight. Glue on numbers from one to ten. Roll a ball to knock the bottles over.

● Picture matching lotto – there are endless combinations to this game. Divide eight A4 sheets of card into six. Find pictures of toys, ensuring that there are two of each. Stick 24 different pictures on four of the cards and the other 24 on the remaining cards. Cut four cards into individual squares. Play with four children and give each a large card. Turn over the small cards and if a child has a matching picture on their large card they cover it.

● Cup and ball – follow the instructions on page 80, using the bottom half of a plastic bottle, string and a paper ball.

● Marble rolling – use an inverted shoe box with arches cut out attached to a tray. Roll the marble in and out of the arches.

● Puppets – see display on page 68.

Talk about

● Compare 'bought' games with the home-made versions. Talk about the differences.

● Look at children's favourite bought games and talk about how they could use a variety of materials to make their own versions.

Home links

● Send home sheets explaining how to make one of the games on display at home.

Further display table ideas

● Make a display of games made by parents and grandparents.

● Have a theme of home-made seasonal games, such as conkers, or games for outdoors.

● Display books and pictures of home-made games from other cultures. Try to make some examples of these with the children.

Daily routines

Colour the pictures and cut them out. Put them in the correct order.

Observation walk

Tick the boxes when you see each object on your walk.

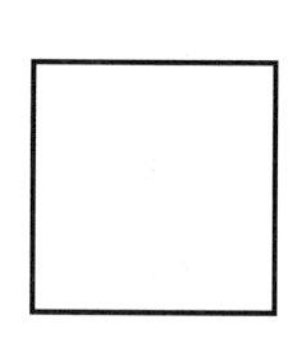

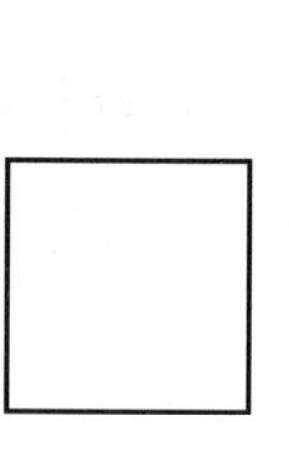

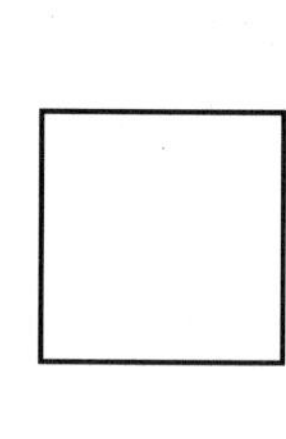

Photocopiable

Perfumed present

Follow the instructions to make a sweet-smelling gift.

What you need

Fine net or kitchen cloth
Lavender or pot pourri
Scissors
Large needle and wool
Thin ribbon

What to do

Cut the net or cloth into a circle.

Sew carefully around the edge of the circle, leaving some spare wool at each end.

Put some lavender or pot pourri in the centre of the circle.

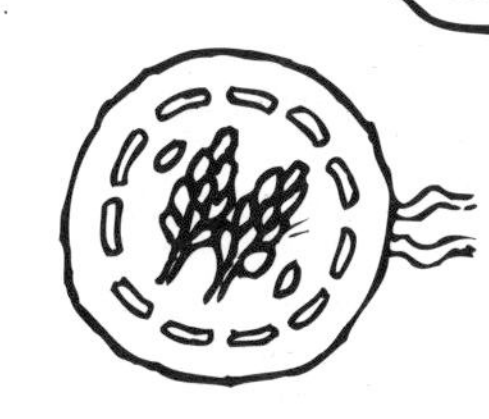

Pull the two loose ends of wool to draw the net or cloth around the contents of the circle.

Tie the ends tightly.

Fasten a piece of ribbon around the bag to form a hanging loop.

Shoe pairs

Draw a line to join the matching pairs of shoes.

Photocopiable

The enormous turnip

Colour in the picture and talk about the words.

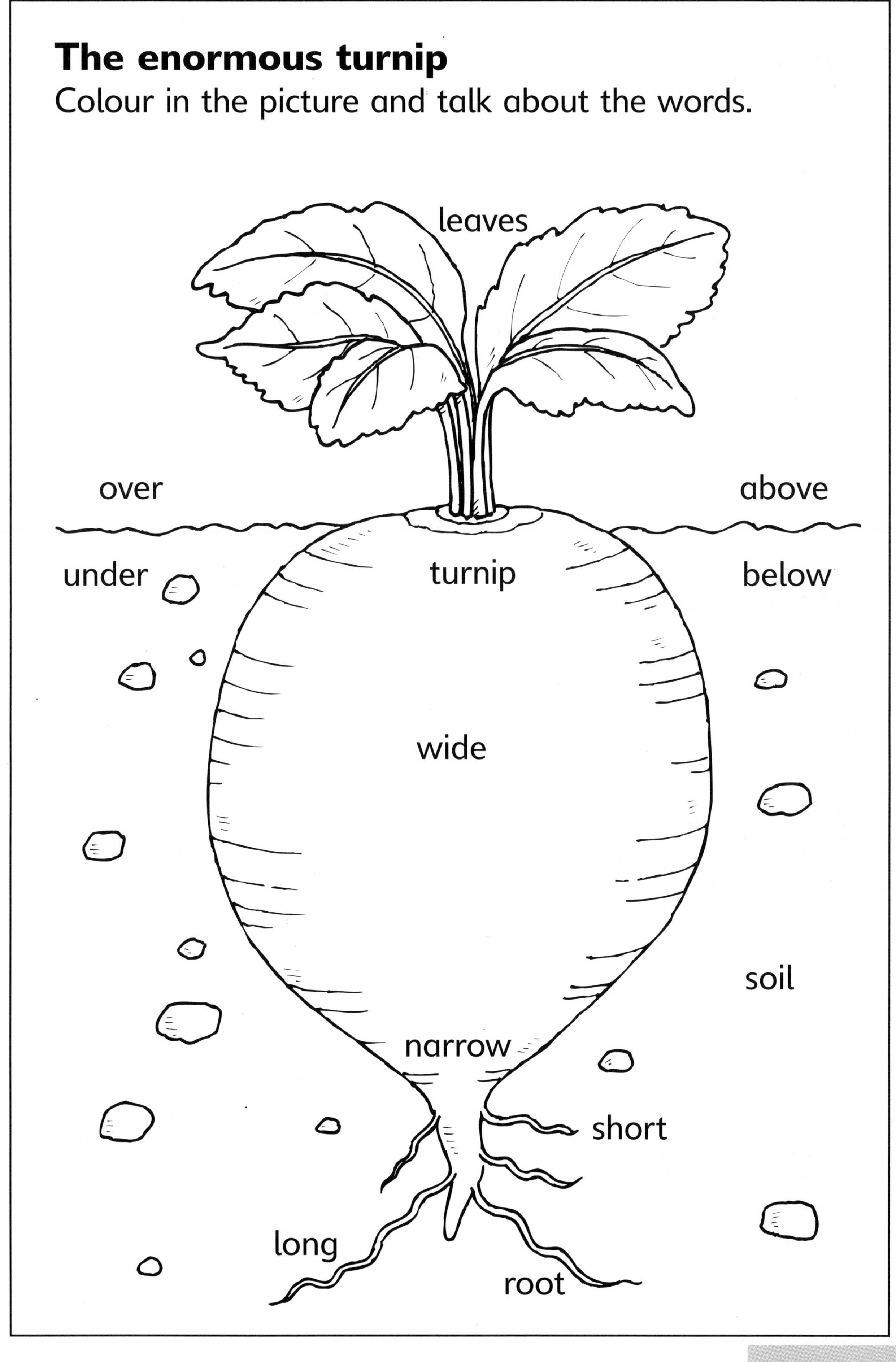

Photocopiable

Salt dough recipe

Follow the recipe to make salt dough models.

What you need

2 cups of flour
1 cup of salt
1 tablespoon of cooking oil
1 cup of water
Powder paint or food colouring if required

What to do

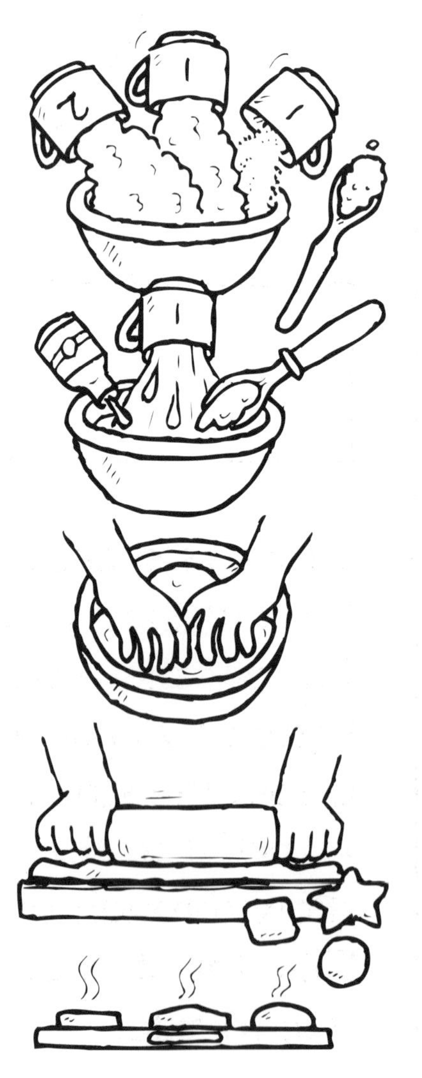

Mix the flour and salt together and add the powder paint if used.

Add the water and oil and a few drops of food colouring if used.

Knead the mixture.

Form shapes using your hands or roll the mixture out and use shape cutters.

Bake in the oven at 180°C or Gas Mark 4 for about an hour.

Moving eyes

Follow the instructions to make a moving eye picture.

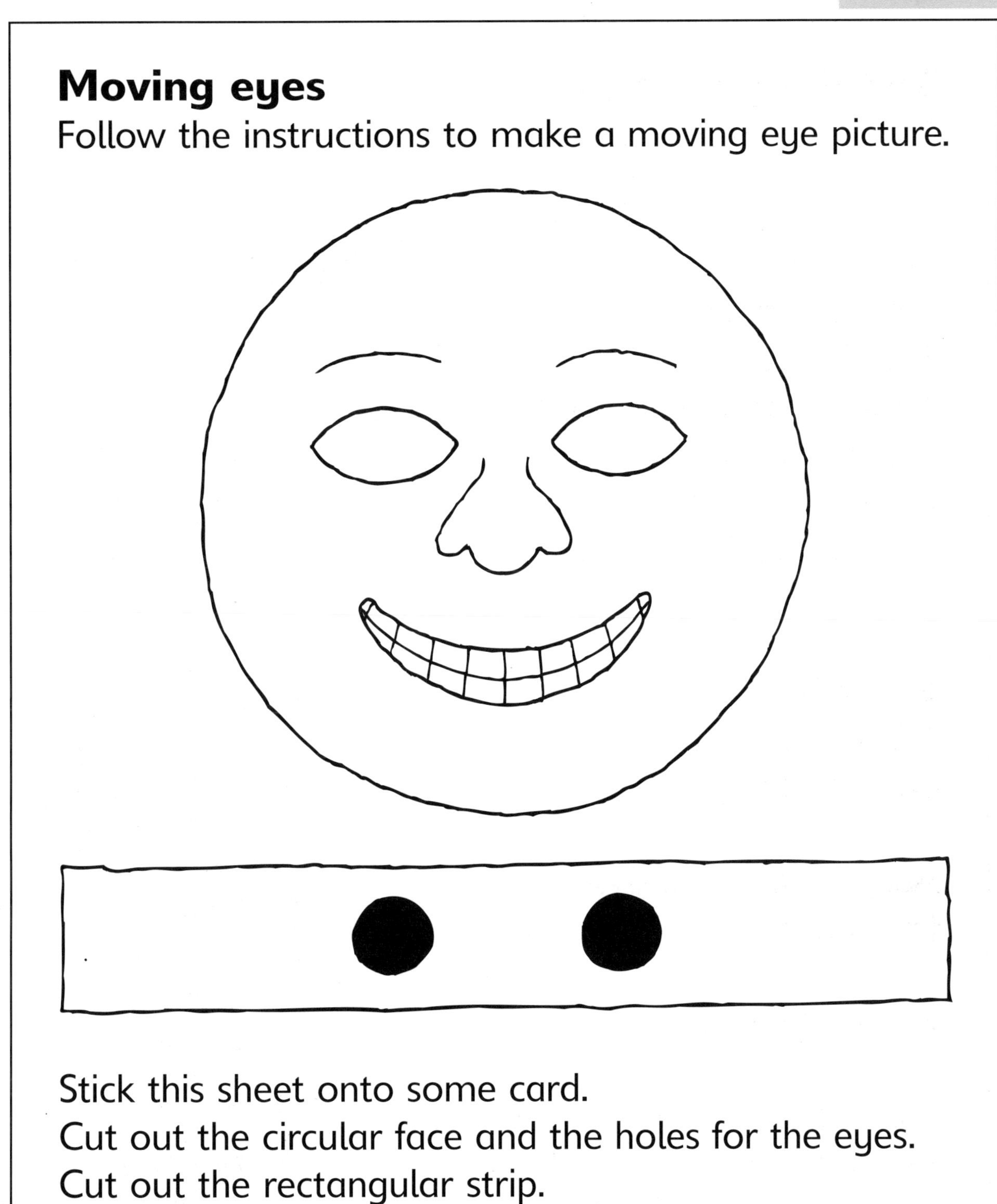

Stick this sheet onto some card.
Cut out the circular face and the holes for the eyes.
Cut out the rectangular strip.
Fasten two pieces of card to the back of the circle, across the strip, to hold it in place.
Attach it to the back of the circle so that the two 'eyes' show through the holes.
Slide the strip of card and watch the eyes move from side to side!

Photocopiable

Cup and ball game

Follow the instructions to make a table-top game.

What you need

Plastic drink bottle
Hole punch
Newspaper
Sticky tape
String

What to do

Cut the plastic bottle in half carefully.

Use the half with the lid and punch a hole in the side near the cut surface.

Screw up some newspaper into a ball and fasten tape around it tightly.

Tie a piece of string through the hole in the bottle and tape it to the ball.

Hold the 'handle' of the bottle and swing it backwards and forwards.

Try to catch the ball in the bottle.